Mono- and Tetranuclear Complexes Featuring (OSSO)- or Tris(ONNO)-Type Ligands for Selective Catalysis

Von der Fakultät für Mathematik, Informatik und Naturwissenschaften der RWTH Aachen University zur Erlangung des akademischen Grades eines Doktors der Naturwissenschaften genehmigte Dissertation

vorgelegt von

Master of Science

Tobias Schindler

aus

Goslar

Berichter: Universitätsprofessor Dr. rer. nat. Jun Okuda
Professor Kazushi Mashima, PhD

Tag der mündlichen Prüfung: 19.09.2019

Denktraditionen im Dialog: Studien zur Befreiung und Interkulturalität, Band 45
Rául Fornet-Betancourt (Hrsg.)

Raúl Fornet-Betancourt

Con la autoridad de la melancolia
Los humanismos y sus melancolias

ISBN: 978-3-95886-306-4

Bibliografische Information der Deutschen Bibliothek
Die Deutsche Bibliothek verzeichnet diese Publikation in der Deutschen Nationalbibliografie; detaillierte bibliografische Daten sind im Internet über http://dnb.ddb.de abrufbar.

Das Werk einschließlich seiner Teile ist urheberrechtlich geschützt. Jede Verwendung ist ohne die Zustimmung des Herausgebers außerhalb der engen Grenzen des Urhebergesetzes unzulässig und strafbar. Das gilt insbesondere für Vervielfältigungen, Übersetzungen, Mikroverfilmungen und die Einspeicherung und Verarbeitung in elektronischen Systemen.

Vertrieb:
1. Auflage 2019
© Verlagshaus Mainz GmbH Aachen
Süsterfeldstr. 83, 52072 Aachen
Tel. 0241/87 34 34 00
Fax 0241/87 55 77
www.Verlag-Mainz.de

Herstellung:
Druckerei Mainz GmbH Aachen
Süsterfeldstraße 83
52072 Aachen
Tel. 0241/87 34 34 00

Satz: nach Druckvorlage des Autors
Umschlaggestaltung: Druckerei Mainz

printed in Germany

The work delineated in this thesis was performed within the International Research Training Group *Selectivity in Chemo- and Biocatalysis* (*SeleCa*) in the laboratories of Univ.-Prof. Dr. Dr. h. c. Jun Okuda at the Institute of Inorganic Chemistry (RWTH Aachen University) and in the laboratories of Prof. PhD Kazushi Mashima at the Graduate School of Engineering Science (Osaka University) between May 2016 and March 2019.

Table of Content

Table of Content i
Table of Abbreviations iii

A.   **General Introduction** 1

    A.1   *Tetradentate Bis(phenolato) (OSSO)- and (ONNO)-Type Ligands* 1
    A.2   *Macrocyclic Tris(ONNO)-Type Ligands* 5
    A.3   *Aim and Scope of this Thesis* 12
    A.4   *References* 13

B.   **Results and Discussion** 17

    B.1   *Group 6 Metal Complexes Featuring a Tetradentate (OSSO)-Type Ligand* 17
        B.1.1   Introduction 17
        B.1.2   Results and Discussion 20
        B.1.3   Summary and Outlook 37
        B.1.4   Experimental 39
        B.1.5   References 44

    B.2   *Mononuclear Complexes Featuring a Tris(ONNO)-Type Ligand* 47
        B.2.1   Introduction 47
        B.2.2   Results and Discussion 51
        B.2.3   Summary and Outlook 67
        B.2.4   Experimental 69
        B.2.5   References 78

    B.3   *Heterometallic Tetranuclear Complexes Featuring a Tris(ONNO)-Type Ligand* 79
        B.3.1   Introduction 79
        B.3.2   Results and Discussion 82
        B.3.3   Summary and Outlook 108
        B.3.4   Experimental 110
        B.3.5   References 117

    B.4   *Catalytic Transformations of Heterocumulenes with Epoxides* 119
        B.4.1   Introduction 119
        B.4.2   Results and Discussion 124
        B.4.3   Summary and Outlook 135
        B.4.4   Experimental 136
        B.4.5   References 138

Table of Content

C.	**Summary**		**141**
D.	**Appendix**		**145**
	D.1	*General Experimental Procedures*	*145*
	D.2	*Crystal Structure Parameters*	*148*
	D.3	*References*	*150*
	D.4	*Table of Compounds*	*151*
	D.1	*Eidesstattliche Erklärung*	*153*
	D.2	*Curriculum Vitae*	*154*
	D.3	*Table of Publications*	*155*
		D.3.1 Peer-Reviewed Publications	155
		D.3.2 Conference Contributions	155
		D.3.3 Other Publications	156
	D.4	*Acknowledgements – Danksagung*	*157*

Table of Abbreviations

A	surface area
A_{iso}	isotropic hyperfine coupling constant
B_0	magnetic field
br	broad signal
c	concentration
$^{C2Ph2}L$	[3+3] macrocyclic ligand featuring a 1,2-diphenylethylene-diamine bridging unit
^{C3}L	[3+3] macrocyclic ligand featuring a 2,2-dimethylpropane-1,3-diamine bridging unit
^{Cy}L	[3+3] macrocyclic ligand featuring a 1,2-cyclohexane-diamine bridging unit
CHC	cyclohexene carbonate
CHO	cyclohexene oxide
COC	cyclic organic carbonates
COSY	correlated spectroscopy
CV	cyclic voltammetry
D_0	diffusion coefficient of analyte
DCM	dichloromethane
DMF	N,N-dimethylformamide
DMSO	dimethyl sulfoxide
DPV	differential pulse voltammetry
$E_{1/2}$	half wave potential
EI	electron ionization
E_P	peak potential
EPR	electron paramagnetic resonance
equiv.	equivalents
ESI	electron-spray-ionization
Et	ethyl
Et$_2$O	diethyl ether
F	Faraday constant
FAB	fast atom bombardment
Fc/Fc$^+$	ferrocene/ferrocenium
g	dimensionless g-factor for characterization of the magnetic moment
GC	gas chromatography
$\hat{H}$	spin Hamiltonian
HSQC	heteronuclear single quantum coherence
i_p	peak current
IR	infrared
J_{ij}	interaction parameter
Ln	lanthanide
I	nuclear spin quantum number
μ_B	Bohr magneton
μ_{eff}	effective magnetic moment without diamagnetic corrections
μ_s	spin-only magnetic moment
m	multiplet
Me	methyl
MeCN	acetonitrile
MeOH	methanol
M_m	molar magnetization
M_n	number average molecular weight
MS	mass spectrometry
M_w	weight average molecular weight
N_A	Avogadro's constant

Table of Abbreviations

^{n}Bu	*n*-butyl
$^nJ_{xy}$	coupling constant
NMO	*N*-methylmorpholine *N*-oxide
NOESY	nuclear Overhauser effect spectroscopy
OAc	acetate
OAT	oxygen atom transfer
(OSSO)	dithiaalkanediyl-2,2'-bis(phenolato) ligand
OTf	triflate
OTs	tosylate
p	pressure
PDI	polydispersity index
Ph	phenyl
PO	propylene oxide
R	ideal gas constant
r.t.	room temperature
RI	refractive index
ROESY	rotating-frame nuclear Overhauser effect correlation spectroscopy
s	singlet
S	spin quantum number
$\hat{S}_n$	spin operator
salalen	amino-imino-alkane bridged bis(phenolate) ligand
salan	diaminoalkane bridged bis(phenolate) ligand
salen	diiminoalkane bridged bis(phenolate) ligand
^{s}Bu	*sec*-butyl
SEC	size exclusion chromatography
SQUID	superconducting quantum interference device
(SSSS)	dithiaalkanediyl-2,2'-bis(thiophenolato) ligand
τ_5	structural parameter for pentacoordinate complexes
T	temperature
^{t}Bu	*tert*-butyl
THF	tetrahydrofuran
TMS	tetramethylsilane
TOF	turnover frequency
UV	ultraviolet
$\tilde{\nu}$	wavenumbers
VT	variable temperature
δ	chemical shift
$v^{1/2}$	square root of the scan rate
χ_m	molar magnetic susceptibility
tris(ONNO)	[3+3] macrocyclic ligand based on three salen-type binding sites
18c6	18-crown-6 ether

A. General Introduction
A.1 Tetradentate Bis(phenolato) (OSSO)- and (ONNO)-Type Ligands

The tetradentate dithiaalkanediyl-2,2'-bis(phenolato), (OSSO)-type, ligand framework was first reported by Wieghardt and co-workers in 2002.[1] Okuda and co-workers further developed the ligand framework for application in olefin polymerization catalysts and prepared ligands with highly tunable properties.[2-7] These (OSSO)-type ligands coordinate to metal centers in L_2X_2-type fashion and feature two hard phenolate (X-type) and two soft thioether (L-type) groups.[8] The ligand system was originally a further development of the related tridentate (OSO)-type ligands by Kakugo and co-workers.[9-16] Theoretical studies on the (OSO)-type ligands indicated interaction between the sulfur donors and the Lewis acidic metal centers, resulting in enhanced polymerization activity.[17, 18]

The properties of the (OSSO)-type ligand framework can be modulated by varying of the bridging units and the *ortho* (R^1) and *para* (R^2) substituents (Figure A.1.1).[2-7] The steric bulk around the metal center can be tuned through different substituents in the *ortho* position (R^1), which also influence the stereorigidity of the complexes: Less bulky groups allow fluxional behavior of the ligand sphere in solution, whereas sterically more demanding substituents produce stereorigid complexes.[4, 19] The rigidity of the ligand framework is also determined by the bridging units: Linear alkanediyl bridges increase the flexibility of the ligand framework whereas cyclic alkanediyl bridges increase its rigidity and introduce chirality into the ligand backbone.[4, 5] Introducing a 1,1'-ferrocenyl bridge adds an additional redox center to the ligand framework that can be used to alter the electronic properties at the metal center, thus enabling redox-switchable catalysis.[20, 21] The electronic properties at the metal center can also be

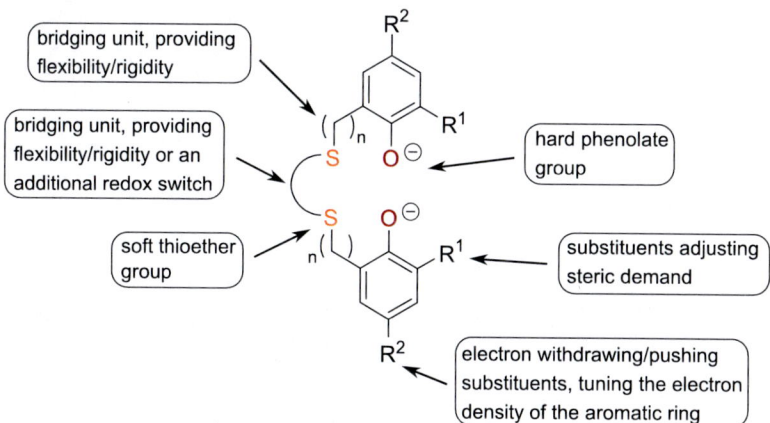

Figure A.1.1. Properties of the tetradentate (OSSO)-type bis(phenolate) ligand framework.

Tetradentate Bis(phenolato) (OSSO)- and (ONNO)-Type Ligands

modulated by introducing electron withdrawing or electron pushing substituents in *para* position (R^2).[4] The different sub-units of the (OSSO)-type ligand framework provide very tunable properties of the metal complexes, demonstrated by the large number of reported pro-ligands.

The hemilabile thioether donors of the (OSSO)-type ligand framework exhibit a weaker interaction to the metal center than the nitrogen donors of the structurally related (ONNO)-type salan, salen and salalen ligands (Figure A.1.2).[2, 22-28] Salan ligands are tetradentate bis(phenoxy)diaminoalkane ligands featuring sp^3 hybridized amine nitrogen atoms, which retain the flexibility of the ligand framework.[29, 30] Salan ligands can be regarded as the nitrogen analogues of the (OSSO)-type ligands (Figure A.1.2(a)). Similar to these ligands, the (ONNO)-type ligand framework is highly modular, and its properties can be tuned by variation of the bridging units and the *ortho* and *para* substituents. An additional modification site is available in the salan ligands as the nitrogen amines can be protonated (R^3 = H) or alkylated (R^3 = alkyl).[31, 32] Salan ligands can be prepared by reduction of their corresponding Schiff bases with LiAlH$_4$ or NaBH$_4$, by reductive amination or Mannich reaction.[30-34] Complexes featuring salan-type ligands have been much less studied than their oxidized salen analogues.[29]

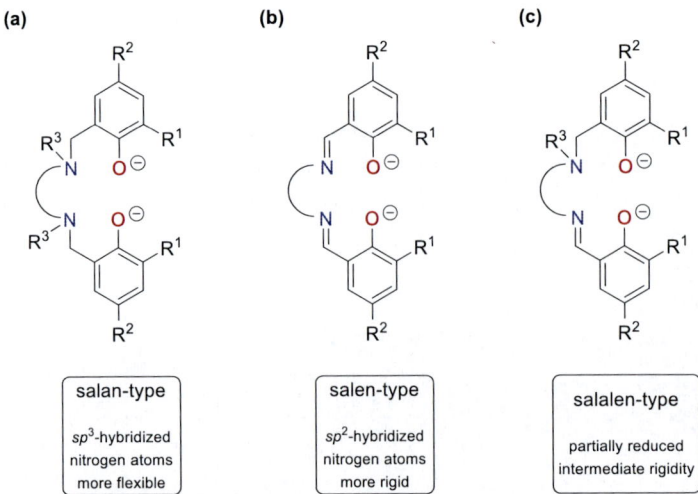

Figure A.1.2. Properties of tetradentate (ONNO)-type (a) salan, (b) salen and (c) salalen ligands.

Salen (ONNO)-type ligands were already prepared in the 1930s through condensation of salicylaldehyde with diamines to produce the corresponding Schiff base.[35] The condensation reaction commonly proceeds without the need of an additional dehydration agent.[29] These ligands are highly modular and a large library of pro-ligands exists, due to the large number of various (commercially) available diamines and hydroxybenzaldehydes and the straightforward preparation of these ligands.[25-27, 29] In contrast to the analogous (OSSO)- and salan (ONNO)-type ligands, salen ligands feature sp^2 hybridized nitrogen atoms, rendering the ligand framework more rigid (Figure A.1.2(b)). A further development of the (ONNO)-type ligands are salalen ligands that feature a partially reduced salen backbone with one sp^3 and one sp^2 hybridized nitrogen donor (Figure A.1.2(c)). Salalen ligands combine the flexibility of salan ligands and the rigidity of salen ligands and, thus, exhibit intermediate stereorigidity.

Due to the highly tunable ligand framework, a large number of complexes featuring (OSSO)- or (ONNO)-type ligands have been reported. Complexation commonly involves the protonolysis of metal precursors with protonated pro-ligands or, alternatively, salt metathesis of deprotonated pro-ligands with metal halides (Scheme A.1.1).[29]

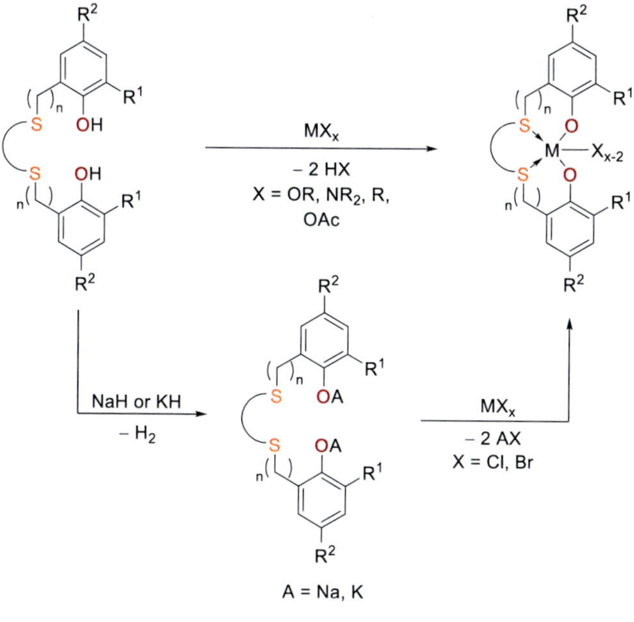

Scheme A.1.1. Complexation reaction through protonolysis of metal precursors with protonated pro-ligands or, alternatively, salt metathesis of deprotonated pro-ligands with metal halides exemplified by (OSSO)-type ligands.

Various coordination geometries have been reported for these ligands (Figure A.1.3): Salen or salan ligands coordinate to tetracoordinate metal centers in square planar fashion. For pentacoordinate metal complexes, salen ligands preferably produce complexes with a square pyramidal coordination geometry in which the oxygen and nitrogen atoms occupy the basal positions, and the additional ligand is situated in the apical position. However, trigonal bipyramidal coordination geometries have also been reported for complexes featuring the more flexible salen ligand with the C_3-bridge propylene-1,3-diamine.[36] Due to the increased flexibility of salan and (OSSO)-type ligands, these ligands produce pentacoordinate complexes with trigonal bipyramidal or square planar coordination geometries.[23, 37] In trigonal bipyramidal coordination geometry, one oxygen and one nitrogen/sulfur atom are located in the apical positions, whereas the other two are situated in the equatorial plane. To determine the degree of trigonality in pentacoordinate complexes, within the continuum between trigonal bipyramidal and square pyramidal, the geometric τ parameter has been defined.[38] The τ_5 parameter can be calculated by comparing the largest angle (β) with the next smaller one (α) with the expression $\tau_5 = (\beta - \alpha)/60°$. A τ parameter with the value of 0 indicates a perfect square pyramidal coordination geometry, whereas a value of 1 indicates a perfect trigonal bipyramidal geometry. In hexacoordinate metal complexes, all three ligands favorably produce complexes with octahedral coordination geometry.

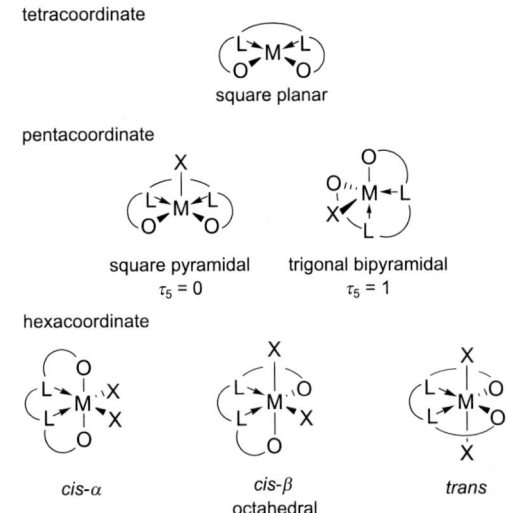

Figure A.1.3: Possible coordination modes of tetra-, penta- or hexacoordinate metal complexes featuring bis(phenolato) salan, salen or (OSSO)-type ligands.

In octahedral complexes, salen ligands predominantly adopt *trans* geometry in which the oxygen and nitrogen atoms are located in the equatorial positions, due to the rigid imine backbone.[29] For more flexible bridges, *cis-β* and *cis-α* coordination of the salen ligand may also occur, although the latter one tends to be the least favorable coordination mode. Salan and (OSSO)-type ligands predominantly adopt helical *cis-α* coordination of the metal center to produce a chiral C_2-symmetric complex, in which the two oxygen atoms are located in the

General Introduction

apical positions and the sulfur atoms in the equatorial plane.[2-4, 9] In some complexes, salan and (OSSO)-type ligands may also coordinate in *cis-β* or *trans* geometry to the metal center.[24, 29] Salalen ligands are arranged in in *cis-β* coordination to afford a chiral C_1-symmetric complex.[29] Here, one oxygen atom is located in the apical position, whereas the nitrogen atoms and the other oxygen atom are located in the equatorial plane. Interconversion between the different coordination geometries may occur in more flexible ligand frameworks although it is inhibited in more rigid ones.

Cis-α coordination of the ligands produces the respective Δ- and Λ-enantiomers of the complexes due to helical chirality (Scheme A.1.2).[3] Complexes featuring flexible (OSSO)-type ligand backbones interconvert between the two enantiomers, exhibiting a fluxional behavior in solution.[3, 4] Interconversion between the Δ- and Λ-enantiomers rapidly occurs in solution in case the ligand backbone has small *ortho* substituents.[19] Larger *ortho* substituents inhibit the exchange process and stabilize the chiral configuration in solution.[4] Due to the high flexibility of the C_3-bridging unit 1,3-propanediyl, complexes with this bridging unit rapidly interconvert in solution at lower temperatures even with bulky *tert*-butyl substituents in *ortho* position.[4] In case of salen and salan ligands, the nitrogen donors exhibit stronger interaction to the metal center and, thus, prevent a fluxional behavior in solution.[39]

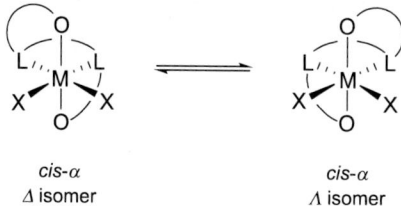

cis-α
Δ isomer

cis-α
Λ isomer

Scheme A.1.2. Interconversion of the Δ- and Λ-enantiomers in *cis-α* coordinated octahedral complexes.[3, 4]

A.2 Macrocyclic Tris(ONNO)-Type Ligands

A further development of salen (ONNO)-type ligands are macrocyclic ligands consisting of two or more salen-type subunits.[40] These ligands can be generally synthesized through condensation of dialdehydes with suitable diamines.[40-42] Macrocyclic ligands have been successfully employed in the preparation of mixed multinuclear complexes due to the availability of different coordination sites.[43-49] When the (ONNO)-subunits form a complex, the phenolic groups are deprotonated to produce more polarized metal-phenolate bonds. Owing to the higher negative charge at the phenolato oxygen atoms, these groups can further stabilize additional metal centers such as alkali metal, alkaline earth metal, rare earth metal and transition metal ions, affording multinuclear metal complexes.[43-52] Multimetallic complexes have received increasing attention due to their potential application in catalysis, small molecule sensing and their unique magnetic and electronic properties.[40, 53-55] Because of the close proximity of the metal centers, multimetallic complexes are interesting model systems for the study of metallic cooperativity, which is frequently encountered in metalloenzymatic active sites.[53]

The first tris(ONNO)-type macrocycle based on 3,6-diformylcatechol was reported by Nabeshima and co-workers in 2001.[56] The research groups of Nabeshima,[40, 43, 51, 56] MacLachlan[42, 50, 57-60] and Brooker[47, 61-63] have decisively contributed to the field of tris(ONNO)-type ligands. The three salen-type (ONNO) subunits of this ligand scaffold can coordinate to smaller transition metal cations similar to the analogous tetradentate salen ligands. Since neighboring (ONNO) subunits share the same benzene backbone, the six neighboring phenolic groups produce a negatively polarized cavity.[40] This cavity is similar to the one of 18-crown-6 (18c6) and may accommodate larger metal ions. Since the ligand framework is highly conjugated, the macrocyclic ring and, in particular, the (ONNO) subunits are more rigid than the analogous tetradentate salen ligands. In addition to the 18c6- and the three salen-type coordination sites, the macrocyclic ring features catechol subunits. These subunits may be potentially redox non-innocent, since catechols can be readily oxidized to semiquinones and, subsequently, quinones in an overall two-electron oxidation reaction.[64] However, the redox properties of heterometallic complexes featuring these tris(ONNO)-type ligands have not yet been determined. Redox non-innocent ligands are of interest for catalytic redox reactions.[65]

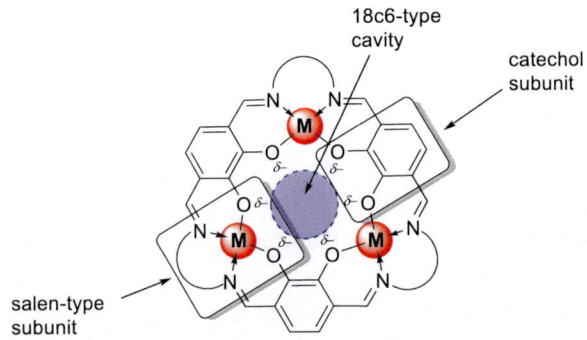

Figure A.2.1. Schematic representation of the respective subunits of the tris(ONNO) macrocycle.[40, 64]

The tris(ONNO) macrocyclic scaffold combines the metal centers in close proximity to each other, producing coordination motives which may allow for cooperative effects of the metal centers applicable in catalytic reactions.[40, 53] Synthesis of the first tris(ONNO)-type pro-ligand was achieved by condensation of three equivalents of 3,6-diformylcatechol with equimolar amounts of benzene-1,2-diamine (Scheme A.2.1, R = H) to produce the corresponding Schiff base.[56] Owing to the rigid aromatic backbone and hydrogen bonding between the phenolic groups and the imine nitrogen atoms, the two chain ends of the linear [3+3]-adduct are conformationally close to each other, facilitating the final condensation step between the terminal aldehyde and amine.[42, 56] Due to π-π-stacking, the compound precipitates from solution, shifting the equilibrium toward the cyclization product and inhibiting side reactions such as oligomerization.[42] However, π-π-stacking also decreases the solubility of the macrocyclic pro-ligand, rendering it unsuitable for application as a ligand of mixed multimetallic complexes.[56] To enhance the solubility of the pro-ligand, different substituents were introduced to the aromatic backbone, producing well-soluble macrocycles based on 4,5-dialkoxybenzene-1,2-diamines.[42, 59]

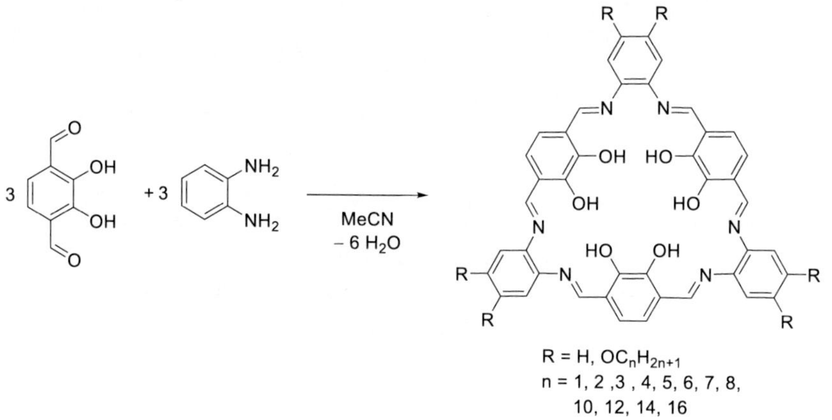

$R = H, OC_nH_{2n+1}$
$n = 1, 2, 3, 4, 5, 6, 7, 8, 10, 12, 14, 16$

Scheme A.2.1. Direct synthesis of the tris(ONNO) macrocycle through a [3+3] cyclization reaction.[40-42]

However, direct synthesis of tris(ONNO) macrocyclic pro-ligands is limited to rigid benzene-1,2-diamines and prone to side reactions. In case of more flexible bridging units such as 1,2-bis(aminooxy)ethane, various oligomeric products are generated upon condensation of the diamine with 3,6-diformylcatechol: Size exclusion chromatography (SEC) indicates formation of [3+3] and [2+2] cyclization products and linear [3+2] and [2+1] oligomers.[43] Formation of such different oligomeric species renders the direct synthesis of the macrocyclic pro-ligands with different bridging units unsuitable for selective pro-ligand synthesis. Monoreduced macrocycles are also commonly generated as a side product during the synthesis of tris(ONNO) pro-ligands (Scheme A.2.2).[57] Condensation of 3,6-diformylcatechol with 4,5-dialkoxybenzene-1,2-diamines produces a benzimidazoline intermediate that reduces the [3+3] macrocycle to generate a monoreduced macrocycle and benzimidazole in a side reaction. The reduction is promoted by residual amounts of acids in commercially available solvents, such as chloroform, and affords the monoreduced macrocycle in moderate to good yields.

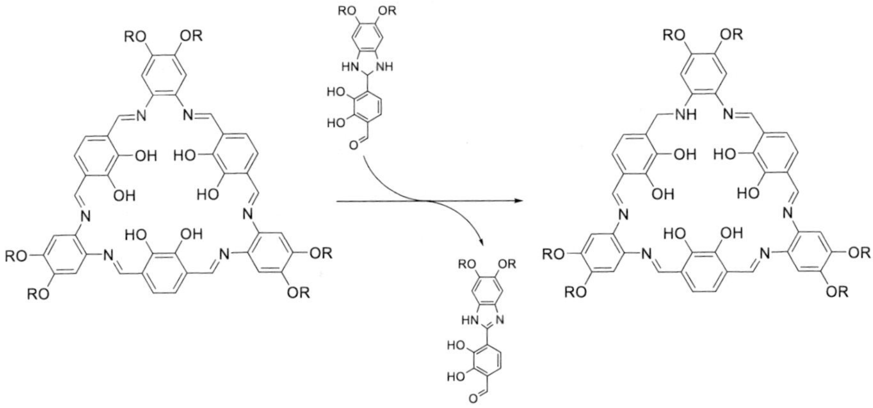

Scheme A.2.2. Reduction of the conjugated macrocycle with benzimidazoline to produce the monoreduced macrocycle and benzimidazole.[57]

To inhibit these side reactions, a core/shell template synthesis was developed by Nabeshima and co-workers (Scheme A.2.3).[43] This approach utilizes the affinity of the respective coordination sites of the macrocycle toward different metal cations to direct the cyclization reaction. Here, 3,6-diformylcatechol pre-coordinates to three small zinc and one large lanthanum cation, preforming the triangular motif of the final macrocycle. In a second step, a cyclization reaction between the dialdehyde units of the lanthanum trizinc cluster and the diamine affords the [3+3] macrocycle. In the resulting lanthanum trizinc complex, the three small zinc cations are exclusively located in the salen-type binding sites, whereas the larger lanthanum cation is exclusively situated in the 18c6-type cavity. In addition to the heterometallic template directed synthesis, an excess of zinc cations may be used to facilitate the [3+3] cyclization reaction.[43] Here, three zinc cations are located in the salen-type subunits and a $[Zn_3(\mu_3\text{-}OH)]^{5+}$ cluster is generated in the 18c6-cavity to produce a hexanuclear complex featuring the tris(ONNO) ligand. During the cyclization reaction a smaller amount of the [4+4] macrocycle is generated as a side product.

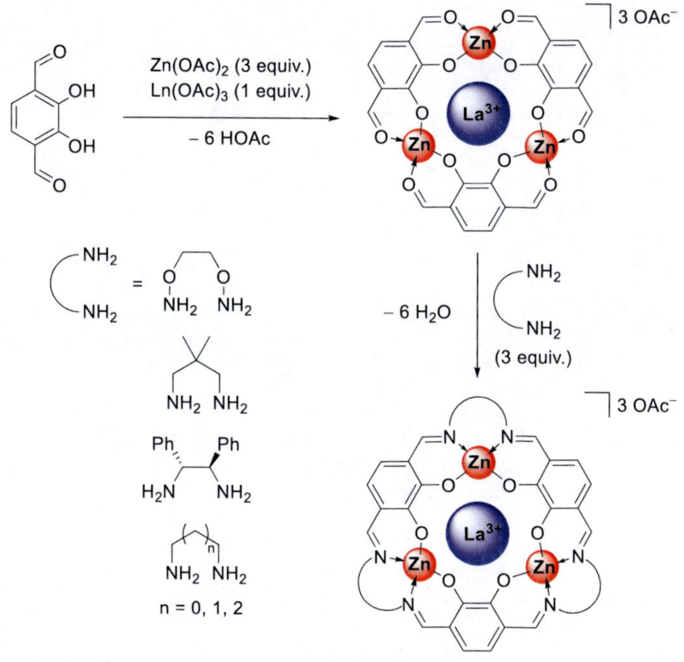

Scheme A.2.3. Core/shell template directed synthesis to prepare tris(ONNO) macrocyclic ligands featuring various bridging units.[43, 44, 46, 48, 49, 66] The template directed synthesis is exemplified by the formation of a lanthanum trizinc complex.

The template directed synthesis has been further developed to produce well soluble macrocycles with substituted or unsubstituted alkanediamine bridging units.[44, 46, 48, 49, 66] The non-coordinating tris(ONNO) pro-ligands can be isolated through removal of the templating metal cations with diluted hydrochloric acid.[43] Due to the labile C=N bonds, template removal is only successful with macrocycles featuring inert oxime bonds which inhibit C=N bond recombinations.[43, 67] In case other bridging units are employed, C=N bond recombinations cleave the [3+3] macrocyclic scaffold and generate various oligomeric Schiff bases. Consequently, protonated tris(ONNO) pro-ligands cannot be isolated in all cases. An overview of all prepared protonated pro-ligands and deprotonated ligands is given in Figure A.2.2.

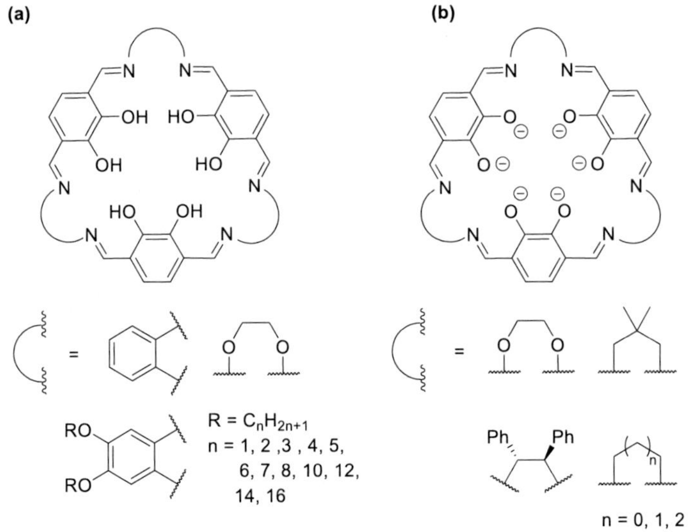

Macrocyclic Tris(ONNO)-Type Ligands

Figure A.2.2. Overview of isolated tris(ONNO) (a) protonated pro-ligands and (b) deprotonated ligands. [40-44, 46, 48, 49, 66]

Complexes featuring tris(ONNO) ligands have been prepared through template directed synthesis or protonolysis of suitable metal precursors with protonated pro-ligands. Depending on the method employed in the preparation of multinuclear complexes, various coordination motives can be achieved (Figure A.2.3). Protonolysis of $M(OAc)_2$ (M = Cu, Co, Ni, Mn, Zn) with the respective protonated pro-ligands affords trinuclear metal complexes with exclusive location of the transition metal cations in the salen-type binding sites (Figure A.2.3(a)).[51, 59] Upon adding further equivalents of zinc or nickel cations, the deprotonated catecholate subunits are successively saturated with transition metal cations to produce tetra-, penta- and hexanuclear complexes, containing up to three zinc or nickel cations in the central 18c6-type cavity (Figure A.2.3(b) – (d)).[43, 50] Treating protonated pro-ligands with an excess of transition metal salts $M(OAc)_2$ (M = Zn, Mn, Cd), the macrocycle acts as a template producing tetranuclear clusters which are located in the 18c6-type cavity to afford an overall heptanuclear complex (Figure A.2.3(e)).[51, 52] Heterometallic complexes have been prepared through core/shell template directed synthesis to produce various tetranuclear lanthanide trizinc or tricopper complexes (Figure A.2.3(f)).[43-49]

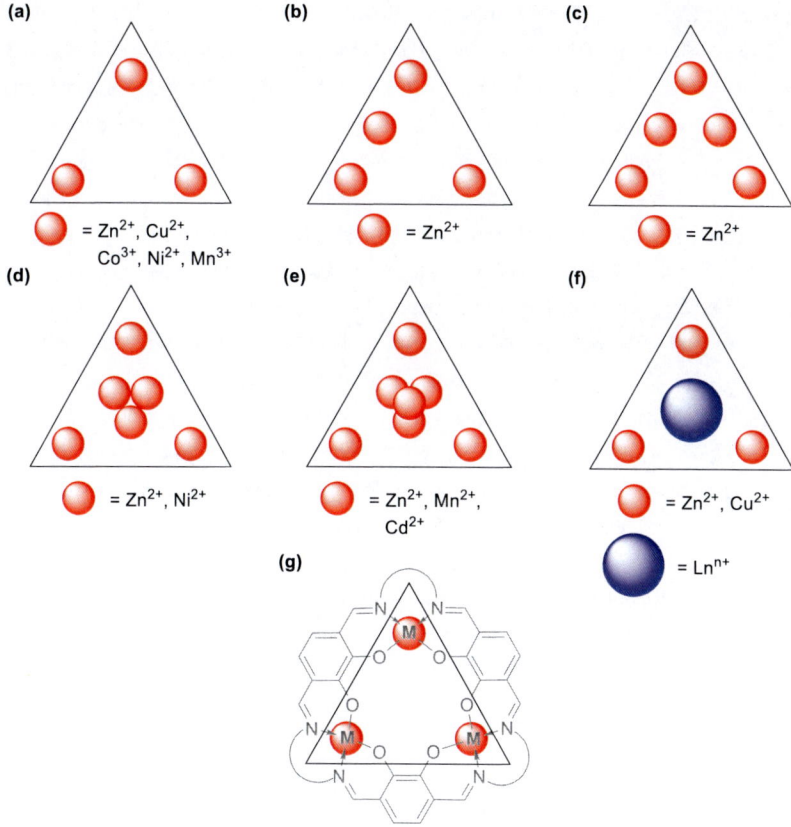

Figure A.2.3. Schematic representation of the various coordination modes of complexes featuring tris(ONNO)-type ligands (TM^{m+} = transition metal, Ln^{n+} = lanthanide): (a) homometallic [L(TM)$_3$]$^{(3m-6)}$ complex,[51, 59] (b) homometallic [L(TM)$_4$]$^{(4m-6)}$ complex,[50] (c) homometallic [L(TM)$_5$]$^{(4m-6)}$ complex.[50], (d) homometallic [L(TM)$_6$]$^{(6m-6)}$ complex with (TM)$_3$$^{3m+}$ cluster in 18c6-cavity,[43, 51] (e) homometallic [L(TM)$_7$]$^{(7m-6)}$ complex with (TM)$_4$$^{4m+}$ cluster in 18c6-cavity,[51, 52] (f) Heterometallic [L(TM)$_3$Ln]$^{(n+m-6)}$ complex,[43-49] and (g) schematic triangular motif of the macrocyclic ligand.

A.3 Aim and Scope of this Thesis

Tetradentate (OSSO)- and (ONNO)-type ligands are structurally related to each other and have been used for stabilizing transition metal complexes.[2-7, 22-28] The high modularity of these ligands makes them interesting for application in complexes featuring redox active transition metals such as molybdenum, vanadium and manganese.[2-7, 29] Whereas (OSSO)-type ligands mostly allow for the preparation of mononuclear transition metal complexes,[2-7] tris(ONNO)-type ligands give access to multinuclear heterometallic complexes featuring the metals in close proximity to each other.[40] This coordination mode may facilitate metallic cooperativity between redox active metal centers that is frequently encountered in metalloenzymatic active sites.[53] The two different binding sites of the tris(ONNO)-type ligands may enable modulation of the redox properties through exchanging the Lewis acidic metal center in the 18c6-type cavity.[53] This thesis examines the structural properties of various metal complexes featuring (OSSO)- or tris(ONNO)-type ligands. The influence of the (OSSO)-type ligand framework onto the redox properties and catalytic performance of molybdenum complexes is elucidated. Since the Lewis acidic metal center in the 18c6-type cavity may alter the properties of heterometallic tris(ONNO)-type complexes, the structural aspects of mononuclear alkaline earth metal and lanthanide metal complexes are explored. The stability of these complexes in subsequent metalation reactions to afford heterometallic trizinc, trivanadium and trimanganese complexes is scrutinized.

Chapter B.1 describes the synthesis and characterization of mononuclear group 6 metal complexes featuring tetradentate (OSSO)-type ligands. The electrochemical and magnetic properties of the complexes are discussed along with their activity in oxygen atom transfer (OAT) reactions of dimethyl sulfoxide (DMSO) with triphenylphosphine (PPh_3).

Chapter B.2 reports the template directed synthesis of mononuclear complexes featuring macrocyclic tris(ONNO)-type ligands. The symmetry in solution and in solid state of these complexes featuring vacant salen-type binding sites is discussed.

In chapter B.3, a new approach toward heterometallic tetranuclear complexes stabilized by tris(ONNO)-type ligands is reported, starting from mononuclear complexes featuring vacant salen-type binding sites. The synthesis of trizinc, trivanadyl and trimanganese complexes is reported along with their structural and magnetic properties.

Chapter B.4 describes the application of mono- and tetranuclear complexes featuring tris(ONNO)-type ligands in catalytic transformations of the heterocumulenes CO_2 and isocyanate with epoxides.

A.4 References

1. Kruse, T.; Weyhermüller, T.; Wieghardt, K. *Inorg. Chim. Acta* **2002**, *331*, 81-89.
2. Capacchione, C.; Proto, A.; Ebeling, H.; Mülhaupt, R.; Möller, K.; Spaniol, T. P.; Okuda, J. *J. Am. Chem. Soc.* **2003**, *125*, 4964-4965.
3. Beckerle, K.; Capacchione, C.; Ebeling, H.; Manivannan, R.; Mülhaupt, R.; Proto, A.; Spaniol, T. P.; Okuda, J. *J. Organomet. Chem.* **2004**, *689*, 4636-4641.
4. Capacchione, C.; Manivannan, R.; Barone, M.; Beckerle, K.; Centore, R.; Oliva, L.; Proto, A.; Tuzi, A.; Spaniol, T. P.; Okuda, J. *Organometallics* **2005**, *24*, 2971-2982.
5. Ma, H.; Spaniol, T. P.; Okuda, J. *Inorg. Chem.* **2008**, *47*, 3328-3339.
6. Meppelder, G.-J. M.; Beckerle, K.; Manivannan, R.; Lian, B.; Raabe, G.; Spaniol, T. P.; Okuda, J. *Chem. Asian J.* **2008**, *3*, 1312-1323.
7. Kapelski, A.; Buffet, J. C.; Spaniol, T. P.; Okuda, J. *Chem. Asian J.* **2012**, *7*, 1320-1330.
8. Green, M. L. H.; Parkin, G. *J. Chem. Educ.* **2014**, *91*, 807-816.
9. Natrajan, L. S.; Wilson, C.; Okuda, J.; Arnold, P. L. *Eur. J. Inorg. Chem.* **2004**, *2004*, 3724-3732.
10. Amor, F.; Fokken, S.; Kleinhenn, T.; Spaniol, T. P.; Okuda, J. *J. Organomet. Chem.* **2001**, *621*, 3-9.
11. Britovsek, G. J. P.; Gibson, V. C.; Wass, D. F. *Angew. Chem. Int. Ed.* **1999**, *38*, 428-447.
12. Okuda, J.; Masoud, E. *Macromol. Chem. Phys.* **1998**, *199*, 543-545.
13. Sernetz, F. G.; Mülhaupt, R.; Fokken, S.; Okuda, J. *Macromolecules* **1997**, *30*, 1562-1569.
14. Fokken, S.; Spaniol, T. P.; Kang, H.-C.; Massa, W.; Okuda, J. *Organometallics* **1996**, *15*, 5069-5072.
15. Miyatake, T.; Mizunuma, K.; Kakugo, M. *Makromol. Chem.-M. Symp.* **1993**, *66*, 203-214.
16. Miyatake, T.; Mizunuma, K.; Seki, Y.; Kakugo, M. *Makromol. Chem.-Rapid* **1989**, *10*, 349-352.
17. Froese, R. D. J.; Musaev, D. G.; Morokuma, K. *Organometallics* **1999**, *18*, 373-379.
18. Froese, R. D. J.; Musaev, D. G.; Matsubara, T.; Morokuma, K. *J. Am. Chem. Soc.* **1997**, *119*, 7190-7196.
19. Snell, A.; Kehr, G.; Wibbeling, B.; Fröhlich, R.; Erker, G. *Z. Naturforsch.* **2003**, *85b*, 838-842.
20. Hermans, C.; Rong, W.; Spaniol, T. P.; Okuda, J. *Dalton Trans.* **2016**, *45*, 8127-8133.
21. Sauer, A. Early Transition Metal Bis(phenolate) Complexes for Selective Catalysis. Doctoral Thesis, RWTH Aachen University, Aachen, 2013.
22. Sauer, A.; Kapelski, A.; Fliedel, C.; Dagorne, S.; Kol, M.; Okuda, J. *Dalton Trans.* **2013**, *42*, 9007-9023.
23. Ma, H.; Melillo, G.; Oliva, L.; Spaniol, T. P.; Englert, U.; Okuda, J. *Dalton Trans.* **2005**, *0*, 721-727.
24. Ma, H.; Spaniol, T. P.; Okuda, J. *Dalton Trans.* **2003**, *0*, 4770-4780.
25. Atwood, D. A.; Harvey, M. J. *Chem. Rev.* **2001**, *101*, 37-52.
26. Cozzi, P. G. *Chem. Soc. Rev.* **2004**, *33*, 410-421.
27. Katsuki, T. *Chem. Soc. Rev.* **2004**, *33*, 437-444.
28. Baleizão, C.; Garcia, H. *Chem. Rev.* **2006**, *106*, 3987-4043.
29. Bellemin-Laponnaz, S.; Dagorne, S., Coordination Chemistry and Applications of Salen, Salan and Salalen Metal Complexes. In *PATAI'S Chemistry of Functional Groups*, Rappoport, Z., Ed. John Wiley & Sons, Ltd.: 2012.
30. Atwood, D. A.; Benson, J.; Jegier, J. A.; Lindholm, N. F.; Martin, K. J.; Pitura, R. J.; Rutherford, D. *Main Group Chem.* **1995**, *1*, 99-113.
31. Subramanian, P.; Spence, J. T.; Ortega, R.; Enemark, J. H. *Inorg. Chem.* **1984**, *23*, 2564-2572.
32. Djebbar-Sid, S.; Benali-Baitich, O.; Deloume, J. P. *Polyhedron* **1997**, *16*, 2175-2182.
33. Tshuva, E. Y.; Gendeziuk, N.; Kol, M. *Tetrahedron Lett.* **2001**, *42*, 6405-6407.

References

34. Hinshaw, C. J.; Peng, G.; Singh, R.; Spence, J. T.; Enemark, J. H.; Bruck, M.; Kristofzski, J.; Merbs, S. L.; Ortega, R. B.; Wexler, P. A. *Inorg. Chem.* **1989**, *28*, 4483-4491.
35. Pfeiffer, P.; Breith, E.; Lübbe, E.; Tsumaki, T. *Liebigs Ann. Chem.* **1933**, *503*, 84-130.
36. Munoz-Hernandez, M.-A.; Keizer, T. S.; Wei, P.; Parkin, S.; Atwood, D. A. *Inorg. Chem.* **2001**, *40*, 6782-6787.
37. Peckermann, I.; Kapelski, A.; Spaniol, T. P.; Okuda, J. *Inorg. Chem.* **2009**, *48*, 5526-5534.
38. Addison, A. W.; Rao, T. N.; Reedijk, J.; van Rijn, J.; Verschoor, G. C. *J. Chem. Soc., Dalton Trans.* **1984**, 1349-1356.
39. Tshuva, E. Y.; Goldberg, I.; Kol, M. *J. Am. Chem. Soc.* **2000**, *122*, 10706-10707.
40. Akine, S.; Nabeshima, T. *Dalton Trans.* **2009**, *47*, 10395-10408.
41. Akine, S.; Taniguchi, T.; Nabeshima, T. *Tetrahedron Lett.* **2001**, *42*, 8861-8864.
42. Gallant, A. J.; Hui, J. K. H.; Zahariev, F. E.; Wang, Y. A.; MacLachlan, M. J. *J. Org. Chem.* **2005**, *70*, 7936-7946.
43. Akine, S.; Sunaga, S.; Taniguchi, T.; Miyazaki, H.; Nabeshima, T. *Inorg. Chem.* **2007**, *46*, 2959-2961.
44. Dhers, S.; Feltham, H. L. C.; Rouzières, M.; Clérac, R.; Brooker, S. *Dalton Trans.* **2016**, *45*, 18089-18093.
45. Feltham, H. L. C.; Clérac, R.; Powell, A. K.; Brooker, S. *Inorg. Chem.* **2011**, *50*, 4232-4234.
46. Feltham, H. L. C.; Clérac, R.; Ungur, L.; Chibotaru, L. F.; Powell, A. K.; Brooker, S. *Inorg. Chem.* **2013**, *52*, 3236-3240.
47. Feltham, H. L. C.; Clérac, R.; Ungur, L.; Vieru, V.; Chibotaru, L. F.; Powell, A. K.; Brooker, S. *Inorg. Chem.* **2012**, *51*, 10603-10612.
48. Feltham, H. L. C.; Lan, Y.; Klöwer, F.; Ungur, L.; Chibotaru, L. F.; Powell, A. K.; Brooker, S. *Chem. Eur. J.* **2011**, *17*, 4362-4365.
49. Nagae, H.; Aoki, R.; Akutagawa, S. N.; Kleemann, J.; Tagawa, R.; Schindler, T.; Choi, J.; Spaniol, T. P.; Tsurugi, H.; Okuda, J.; Mashima, K. *Angew. Chem. Int. Ed.* **2018**, *57*, 2492-2496.
50. Frischmann, P. D.; Gallant, A. J.; Chong, J. H.; MacLachlan, M. J. *Inorg. Chem.* **2008**, *47*, 101-112.
51. Nabeshima, T.; Miyazaki, H.; Iwasaki, A.; Akine, S.; Saiki, T.; Ikeda, C. *Tetrahedron* **2007**, *63*, 3328-3333.
52. Frischmann, P. D.; MacLachlan, M. J. *Chem. Commun.* **2007**, *0*, 4480-4482.
53. Clarke, R. M.; Storr, T. *Dalton Trans.* **2014**, *43*, 9380-9391.
54. Akine, S.; Utsuno, F.; Piao, S.; Orita, H.; Tsuzuki, S.; Nabeshima, T. *Inorg. Chem.* **2016**, *55*, 810-821.
55. Pokharel, U. R.; Fronczek, F. R.; Maverick, A. W. *Nat. Commun.* **2014**, *5*, 5883-5887.
56. Akine, S.; Taniguchi, T.; Nabeshima, T. *Tetrahedron Lett.* **2001**, *42*, 8861-8864.
57. MacLachlan, M. J. *Pure Appl. Chem.* **2006**, *78*, 873-888.
58. Gallant, A. J.; Yun, M.; Sauer, M.; Yeung, C. S.; MacLachlan, M. J. *Org. Lett.* **2005**, *7*, 4827-4830.
59. Gallant, A. J.; Chong, J. H.; MacLachlan, M. J. *Inorg. Chem.* **2006**, *45*, 5248-5250.
60. Frischmann, P. D.; Jiang, J.; Hui, J. K. H.; Grzybowski, J. J.; MacLachlan, M. J. *Org. Lett.* **2008**, *10*, 1255-1258.
61. Feltham, H. L. C.; Brooker, S. *Coord. Chem. Rev.* **2009**, *253*, 1458-1475.
62. Feltham, H. L.; Clerac, R.; Ungur, L.; Chibotaru, L. F.; Powell, A. K.; Brooker, S. *Inorg. Chem.* **2013**, *52*, 3236-3240.
63. Feltham, H. L. C.; Dhers, S.; Rouzières, M.; Clérac, R.; Powell, A. K.; Brooker, S. *Inorg. Chem. Front.* **2015**, *2*, 982-990.
64. Ulas, G.; Lemmin, T.; Wu, Y.; Gassner, G. T.; DeGrado, W. F. *Nat. Chem.* **2016**, *8*, 354-359.
65. Lyaskovskyy, V.; de Bruin, B. *ACS Catal.* **2012**, *2*, 270-279.
66. Yamashita, A.; Watanabe, A.; Akine, S.; Nabeshima, T.; Nakano, K.; Yamamura, T.; Kajiwara, T. *Angew. Chem. Int. Ed.* **2011**, *50*, 4016-4019.

67. Akine, S.; Taniguchi, T.; Dong, W.; Masubuchi, S.; Nabeshima, T. *J. Org. Chem.* **2005,** *70*, 1704-1711.

References

B. Results and Discussion

B.1 Group 6 Metal Complexes Featuring a Tetradentate (OSSO)-Type Ligand

Parts of this chapter were published in: Schindler, T.; Sauer, A.; Spaniol, T. P.; Okuda, J. Oxygen Atom Transfer Reactions with Molybdenum Cofactor Model Complexes That Contain a Tetradentate OSSO-Type Bis(phenolato) Ligand. *Organometallics*, **2018**, *37*, 4336–4340.

B.1.1 Introduction

Oxidoreductases containing a molybdenum cofactor play a crucial role in living organisms and catalyze oxygen atom transfer (OAT) reactions.[1,2] Important representatives of these molybdenum containing enzymes are xanthine oxidase, nitrate reductase, aldehyde oxidase, sulfite oxidase, arsenite oxidase, DMSO reductase and formate dehydrogenase. Commonly, the enzymatic active sites stabilize the metal center with molybdopterin units, featuring molybdenum with terminal oxo or sulfide ligands (Scheme B.1.1(a)). The OAT reactions commonly start from an active mono-oxo molybdenum(IV) complex which is oxidized by the substrate to generate a dioxo molybdenum(VI) species (Scheme B.1.1(b)).[3,4] The oxidized complex is stepwise reduced and protonated to afford the starting mono-oxo molybdenum(IV) complex and water, proceeding *via* a molybdenum(V) intermediate. The molybdopterin ligand contains a dithiolene chelate that coordinates to molybdenum. The dithiolene group is easily oxidized in an overall two-electron-oxidation to a 1,2-dithioketone unit, rendering the ligand redox non-innocent.[2] The redox non-innocence of the coordinating ligand is believed to be pivotal to the catalytic performance of natural molybdenum cofactors. The higher homologue tungsten may substitute molybdenum in these enzymes when molybdenum is not bioavailable.[5]

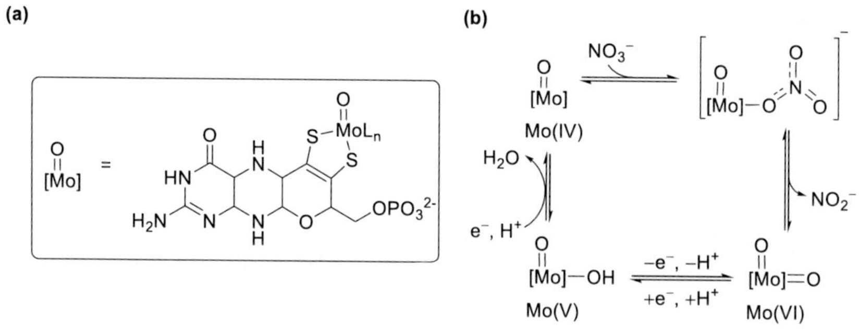

Scheme B.1.1. (a) Molybdenum cofactor stabilized by one molybdopterin unit and (b) catalytic reduction of nitrate to nitrite by molybdenum containing nitrate reductase.[1,2]

Group 6 Metal Complexes Featuring a Tetradentate (OSSO)-Type Ligand

Due to the importance of molybdenum cofactor containing enzymes for living organisms, various chemical model complexes have been prepared since the 1980s to understand and mimic biological OAT reactions.[2, 6] Most model complexes contain mono-oxo molybdenum(IV) or dioxo molybdenum(VI) species coordinated by dithiolate ligands, similar to the ones observed in natural molybdenum oxotransferases (Scheme B.1.2(a)).[7] Due to the complex synthesis of the molybdopterin ligand, a model complex which most closely resembles the molybdenum cofactor active site has been prepared as late as 2012.[8] The key steps of the mechanism have been clarified by spectroscopic and electronic studies on model complexes with molybdenum in the relevant oxidation states +IV, +V and +VI.[2, 9] The reduction of DMSO with physiologically irrelevant phosphines PR_3 (R = Ph, alkyl) commonly serves as a model reaction for the DMSO reductase family (Scheme B.1.2).[1, 2, 10] The investigated OAT reactions commonly start from a bis(oxo) molybdenum(VI) complex which is reduced by phosphine to give a mono-oxo molybdenum(IV) species and phosphine oxide (Scheme B.1.2(b)). As a side reaction, phosphine partially reduces two molybdenum(VI) complexes to produce a μ_2-oxo-bridged molybdenum(V) dimer, rendering the complex inactive.[11] Oxidation of the mono-oxo molybdenum(IV) complex by DMSO regenerates the starting bis(oxo) molybdenum(VI) complex. Contrary to most model systems, the natural reduction of DMSO proceeds via a mono-oxo molybdenum(VI) complex.[2]

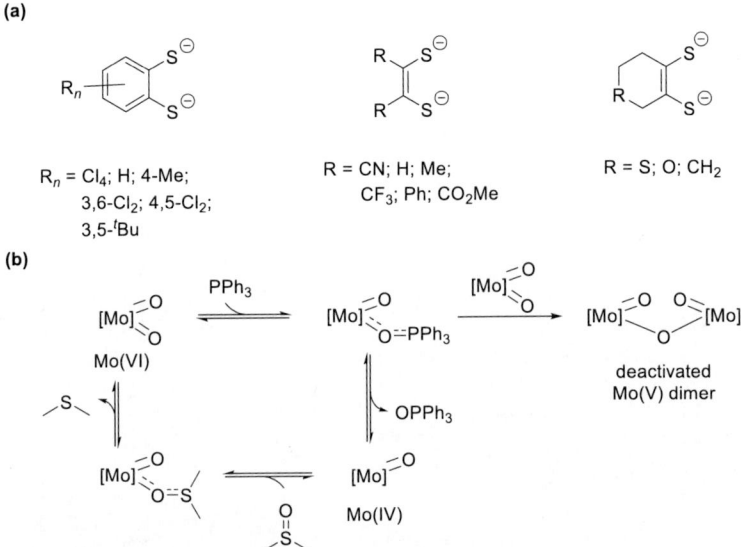

Scheme B.1.2. (a) Common ligands employed in molybdenum cofactor model complexes and (b) catalytic cycle of DMSO reduction with PPh₃ starting from a bis(oxo) molybdenum(VI) complex.[1, 2, 11]

Results and Discussion

Due to the high modularity of tetradentate bis(phenolate) and bis(thiophenolate) ligands, model complexes featuring these ligands have also been prepared (Figure B.1.1). The flexible (ONNO)-, (SNNS)- and (SSSS)-type ligands stabilize high-oxidation-state bis(oxo) molybdenum(VI) centers in cis-α coordination.[11-13] In case of the more rigid salen-type ligand, cis-β coordination of the ligand is observed, rendering one phenolate oxygen atom in apical position and the other one *trans* to a terminal oxo ligand in the equatorial plane (Figure B.1.1).[14] Spectroscopic and electrochemical characterization of [(SSSS)MoO$_2$] and its reduced mono-oxo molybdenum(V) and mono-oxo molybdenum(IV) analogues indicated similar properties to the molybdenum center found in sulfite oxidase.[11]

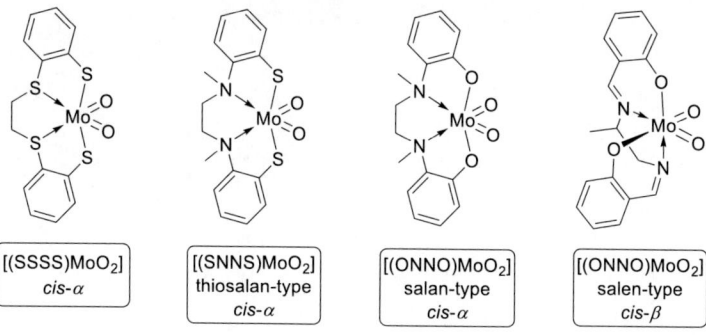

| [(SSSS)MoO$_2$] | [(SNNS)MoO$_2$] | [(ONNO)MoO$_2$] | [(ONNO)MoO$_2$] |
| cis-α | thiosalan-type cis-α | salan-type cis-α | salen-type cis-β |

Figure B.1.1. Bis(oxo) molybdenum(VI) complexes featuring tetradentate bis(thiophenolate) or bis(phenolate) ligands.[11-15]

Okuda and co-workers have recently reported the synthesis and catalytic application of high-oxidation-state bis(oxo) molybdenum complexes [(OSSO)MoO$_2$] featuring (OSSO)-type ligands for deoxydehydration reactions.[16, 17] In a previous study, a bis(chloro) molybdenum(IV) complex was isolated that may serve as a starting complexes for the OAT active mono-oxo molybdenum(IV) complex.[17] Compared to the structurally related (SSSS)-type ligands, the (OSSO)-type ligands feature a hard donor bis(phenolate) ligand, rendering these complexes interesting for comparative studies between the hard and soft donor properties of (OSSO)- and (SSSS)-type ligands.

Group 6 Metal Complexes Featuring a Tetradentate (OSSO)-Type Ligand

B.1.2 Results and Discussion

Inspired by the previous results on molybdenum complexes featuring tetradentate (OSSO)-type ligands,[16, 17] complexes featuring molybdenum or tungsten in the formal oxidation state +IV, +V and +VI were synthesized. Spectroscopic, electrochemical and reactivity studies were performed on these complexes and their catalytic activity in OAT reactions assessed.

B.1.2.1 Characterization and Reactivity Studies of Molybdenum(IV) Complexes

The dichloro molybdenum(IV) complex [(OSSO)MoCl$_2$] (**1a**) was prepared similar to the previously reported complexes **1b** and the analogous [(SSSS)MoCl$_2$].[17-19] **1a** and **1b** were prepared through protonolysis of [MoCl$_4$(NCMe)$_2$] with the respective [(OSSO)H$_2$] pro-ligands (HOC$_6$H$_2$-R^1-4-R^1-6)$_2${S-R^2-S} (R^1 = ^{t}Bu, R^2 = 1,2-butanediyl for **1a**; R^1 = CPhMe$_2$, R^2 = *rac*-1,2-cyclohexanediyl for **1b**) in THF at 60 °C. The complexes were quantitatively obtained as air- and moisture-sensitive orange red solids in good to moderate yields after purification through recrystallization from THF (Scheme B.1.3). Previously the analogous tungsten complexes [(OSSO)WCl$_2$] were reported.[17] However, recent research has indicated that an upscale of the reaction to a preparative scale remains challenging which may be attributed to the lower reactivity of the WCl$_4$ precursor compared to [MoCl$_4$(NCMe)$_2$]. The deprotonated pro-ligands [(OSSO)M$_2$] (M = Li, Na, K) were used to drive the reaction toward completion through salt metathesis. However, the reactions were unselective producing large amounts of insoluble material with undefined composition. Therefore, further research on [(OSSO)WCl$_2$] complexes was discontinued.

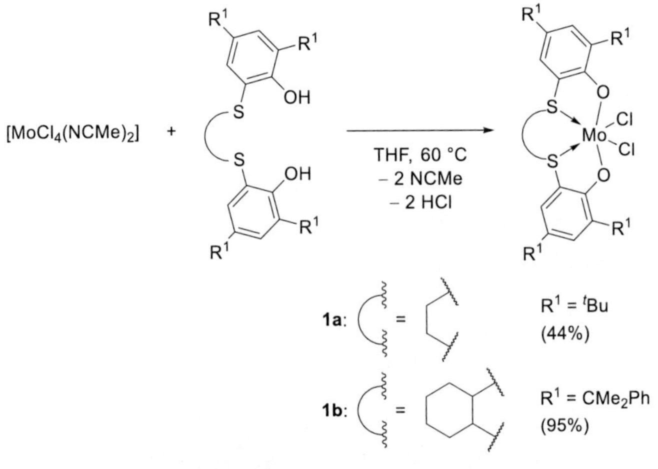

Scheme B.1.3. Synthesis of [(OSSO)MoCl$_2$] (**1a**, **1b**) through protonolysis of (OSSO)H$_2$ with [MoCl$_4$(NCMe)$_2$].[20]

Results and Discussion

Compounds **1a** and **1b** are paramagnetic and exhibit a magnetic moment of μ_{eff} = 2.3 μ_B for **1a** and μ_{eff} = 2.6 μ_B for **1b** in solution (DCM-d_2), which was determined by Evans Method.[21, 22] The magnetic moments are in a typical range for d^2 complexes with a theoretically calculated spin-only magnetic moment of μ_s = 2.8 μ_B for S_{total} = 1 (g ≈ 2).[23] The deviation between the experimentally determined magnetic moments and the spin-only magnetic moment may result from diamagnetic impurities. The high-spin d^2 configuration is rationalized by lacking π-interaction between the *trans*-coordinated phenolate-oxygen and the metal centers, similar to other group 6 high-spin d^2 complexes featuring bulky phenolate ligands.[17, 24] Complexes **1a** and **1b** exhibit large chemical shifts in their respective ^{1}H NMR spectra in DCM-d_2. The temperature dependence of the chemical shifts was investigated to determine whether spin-pairing at lower temperatures occurs through a spin-transition from S_{total} = 1 to S_{total} = 0. ^{1}H NMR spectra were recorded from 293 K to 183 K at an interval of 10 K (Figure B.1.2).

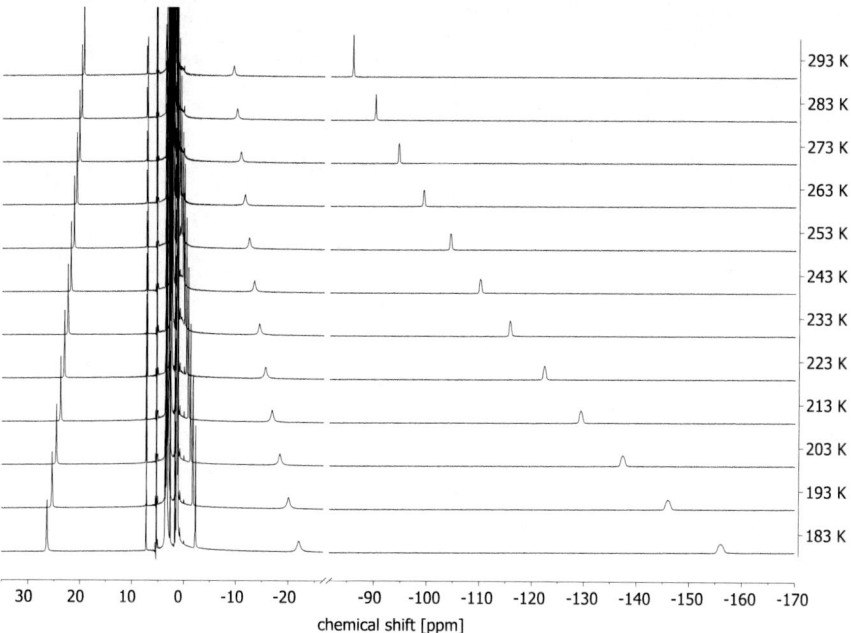

Figure B.1.2. Variable temperature ^{1}H NMR experiment of **1a** in DCM-d_2 in the temperature range from 293K to 183 K.[20]

A large temperature dependence of the chemical shifts was observed: The signal at δ −86 ppm at 298 K is shifted by 70 ppm to δ −156 ppm at 183 K. The chemical shift δ is linearly dependent on the reciprocal temperature (T^{-1}), indicating that the graph follows Curie's law and that the

Group 6 Metal Complexes Featuring a Tetradentate (OSSO)-Type Ligand

compound follows Curie paramagnetism within the evaluated temperature range (Figure B.1.3).[23]

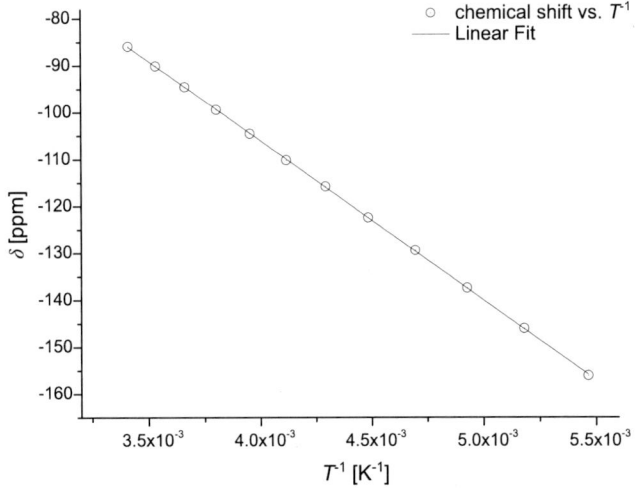

Figure B.1.3. Chemical shift of the signal at δ −85.9 ppm (298 K) plotted against the reciprocal temperature and linear fit of the data to reveal Curie paramagnetism.[20, 23]

The molecular structure of the [(OSSO)MoCl$_2$] complexes was previously established by X-ray diffraction on single crystals of **1b**.[17, 20] The molecular structure of **1b** shows a C_2-symmetric complex with the metal atom coordinated in distorted octahedral geometry. The chlorine atoms are coordinated in *cis*-arrangement in the equatorial plane and the (OSSO) ligand adopts a helical *cis-α* coordination.

To probe whether the reduction of [(OSSO)MoCl$_2$] may lead to selective formation of low-valent molybdenum complexes, the electrochemical properties of **1a** and **1b** were determined by cyclic voltammetry (Figure B.1.4). The cyclic voltammograms of **1a** and **1b** in DCM reveal a reversible oxidation event at $E_{1/2}$ (1) = 0.50 V for **1a** and at $E_{1/2}$ (1) = 0.50 V for **1b**, as well as a quasi-reversible reduction event at $E_{1/2}$ (2) = −1.12 V for **1a** and at $E_{1/2}$ (2) = −0.98 V for **1b** (vs Fc/Fc$^+$, compound (2 mM), electrolyte: [^{n}Bu$_4$N][PF$_6$] (100 mM)). Recording two or more cyclic voltammogram cycles, additional redox events emerged. To further investigate the nature of these redox events, differential pulse voltammograms of **1a** in DCM were recorded (Figure B.1.4). Applying a positive starting potential and stepwise decreasing the potential to negative value, the two expected redox events E_p (1) and E_p (2) were observed (Figure B.1.4c). The absence of additional redox events in the DPV graph in Figure B.1.4c further substantiates the reversible nature of the oxidation event. Recording the differential pulse voltammogram in opposite direction, applying a negative starting potential and increasing the potential to positive value, additional redox events emerged at E_p (3) = −0.25 V and E_p (4) = 0.00 V (Figure B.1.4d). The DPV data suggests that upon complete reduction of **1a** subsequent reactions occur which produce new species. The oxidation of these subsequently generated species account for the redox events E_p (3) and E_p (4). Since the peak currents of E_p (3) and E_p (4) are smaller compared to E_p (1) and E_p (2), the reduced species [**1a**]⁻ is not quantitatively converted to subsequent species, further underlining the quasi-reversible nature of the reduction of **1a**.

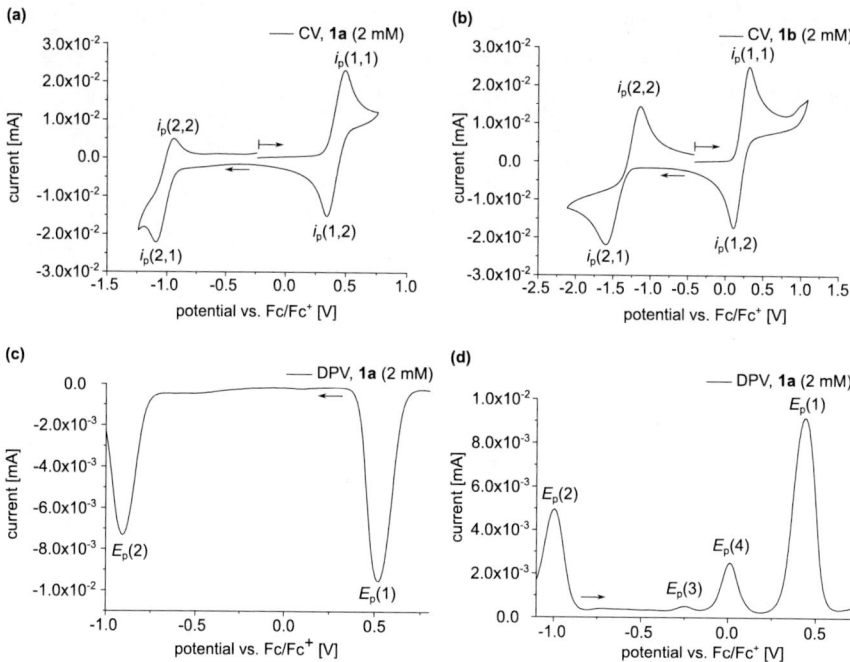

Figure B.1.4. Cyclic voltammograms of (a) **1a** (2 mM) and of (b) **1b** (2 mM), scan rate 100 mV s⁻¹; differential pulse voltammograms of **1a** (2 mM) (c) from positive to negative potential and (d) from negative to positive potential in DCM at 298 K, electrolyte [^{n}Bu₄N][PF₆] (100 mM), referenced to the Fc/Fc⁺ couple.[20]

The reversible and quasi-reversible nature of the redox events was also elucidated through recording cyclic voltammograms at different scan rates (v) (Figure B.1.5). According to the Randless-Sevcik equation (Equation B.1.1), the peak currents (i_p) of electrochemically reversible electron transfer processes with freely diffusing redox species show a linear dependence when plotted against $v^{1/2}$ (n number of electrons transferred, A electrode surface area, D_0 diffusion coefficient of the analyte, c_0 bulk concentration of the analyte, F Faraday constant, R ideal gas constant, T temperature).[25] For such kind of electron transfer processes, the ratios of the peak currents $i_p(1,1)/i_p(1,2)$ and $i_p(2,1)/i_p(2,2)$ are independent of $v^{1/2}$, with zero slope when the ratio of the peak currents are plotted against $v^{1/2}$.

$$i_p = 0.446\, nFAc_0 \left(\frac{nFvD_0}{RT}\right)^{1/2} \quad \text{(Equation B.1.1)}$$

The peak currents of **1a** are linearly dependent on the square root of the scanning rate $v^{1/2}$. The ratios of the peak currents of the oxidative event at $E_{1/2}$ (1) = 0.50 V are distributed on a horizontal line within the margin of error (Figure B.1.5). The distribution of the peak currents on a horizontal line indicates a mostly reversible electron transfer process. On the other hand, the ratios of the peak currents of the reductive event at $E_{1/2}$ (2) = −1.12 V deviate from linearity, thus indicating a quasi-reversible to irreversible electron transfer process.

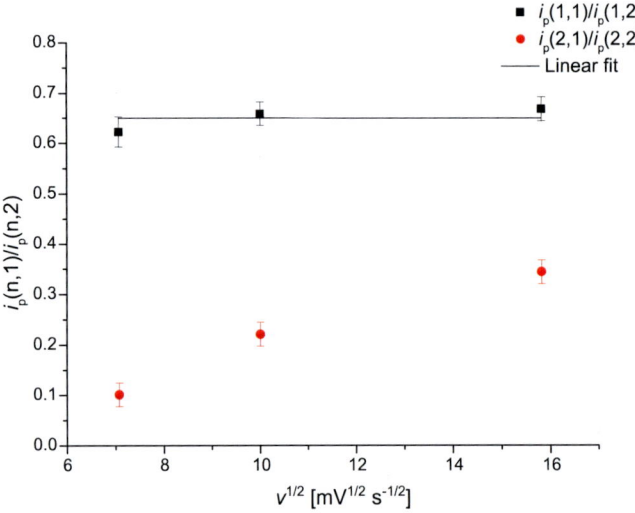

Figure B.1.5. Plot of the ratios of the peak currents (i_p(n,1)/i_p(n,2)) of **1a** against $v^{1/2}$ (v = scan rate).

The redox events of the [(OSSO)MoCl$_2$] complexes can be tentatively assigned to the Mo(IV)/Mo(V) and Mo(IV)/Mo(III) couples. The quasi-reversible reduction may be attributed to dissociation of one chloro ligand and subsequent reactions which have also been reported for similar complexes (Scheme B.1.4).[11]

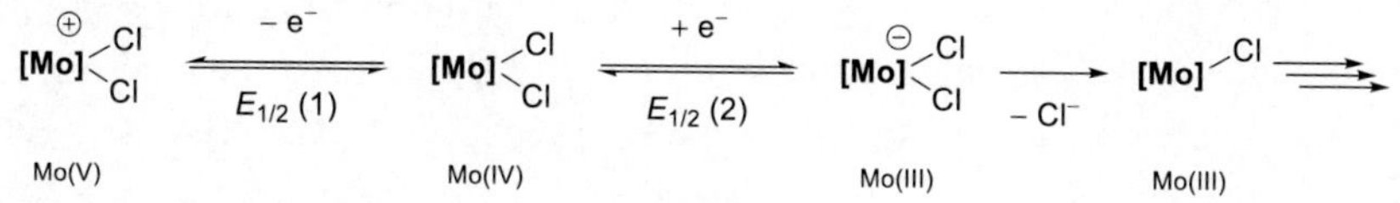

Results and Discussion

$$[Mo]\overset{\oplus}{\underset{Cl}{<}}{}^{Cl} \underset{E_{1/2}(1)}{\overset{-e^-}{\rightleftharpoons}} [Mo]\underset{Cl}{<}{}^{Cl} \underset{E_{1/2}(2)}{\overset{+e^-}{\rightleftharpoons}} [Mo]\overset{\ominus}{\underset{Cl}{<}}{}^{Cl} \xrightarrow{-Cl^-} [Mo]-Cl \rightleftharpoons$$

Mo(V)   Mo(IV)   Mo(III)   Mo(III)

Scheme B.1.4. Electrochemical processes upon oxidation or reduction of the [(OSSO)MoCl$_2$] complexes ([**Mo**] = (OSSO)Mo).

Since the cyclic voltammograms of **1a** and **1b** showed a reversible oxidation, chemical oxidation was anticipated to selectively produce Mo(V) or Mo(VI) complexes. Utilizing oxygen-atom-transfer reactions to introduce oxo-ligands, **1a** was chemically oxidized through addition of DMSO, NMO or N$_2$O to solutions of **1a** in benzene-d_6 (Scheme B.1.5). In case of an excess of NMO or DMSO as oxidant, the ^{1}H NMR spectrum of the reaction showed complete consumption of **1a** in favor of the dioxo complex [(OSSO)MoO$_2$] as well as some [(OSSO)H$_2$] after heating for 16 h at 60 °C.

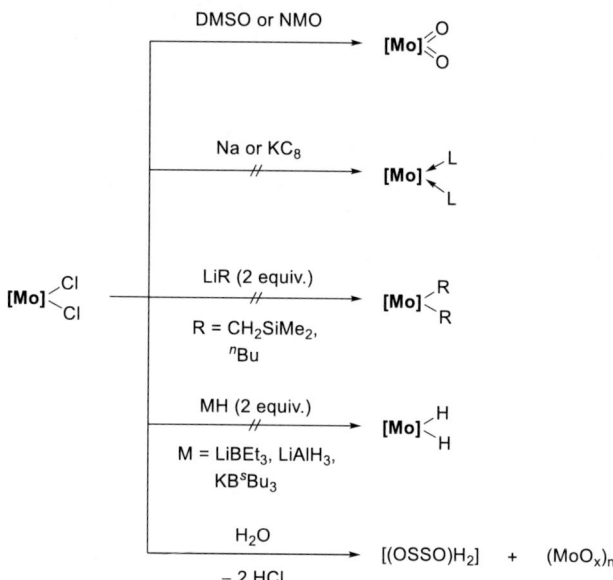

Scheme B.1.5. Reactivity studies of **1a** ([**Mo**] = (OSSO)Mo).

Presumably the intermediate species [(OSSO)MoOCl$_2$] (**A**) is generated upon oxidation which decomposes to [(OSSO)MoOCl] (**2a**) and eventually to [(OSSO)MoO$_2$] (**3**) (refer to section B.1.2.4). In case of N$_2$O no reaction was observed. Attempts to stop the oxidation at the proposed [(OSSO)MoOCl$_2$] intermediate were unsuccessful. Since **1a** was readily oxidized with DMSO, the dichloro Mo(IV) complex was identified as a potential candidate for catalytic OAT reactions (refer to section B.1.2.4). Attempts to produce low-valent Mo(II) complexes through chemical reduction of **1a** with Na or KC_8 were unsuccessful (Scheme B.1.5). Hydrolysis of **1a** with water in presence or absence of a base, such as NEt_3 or DBU, was used to introduce a terminal oxo ligand and prepare a mono-oxo molybdenum(IV) complex (Scheme B.1.5). However, hydrolysis only resulted in formation of the protonated [(OSSO)H_2] pro-ligand together with insoluble solids. Ligand exchange reactions with $LiCH_2SiMe_3$, ^{n}BuLi, Li[HBEt$_3$], K[HB^{s}Bu$_3$], Na[BH$_4$] or LiAlH$_4$ to produce [(OSSO)MoR$_2$] (R = CH_2SiMe_3, ^{n}Bu, H) were unselective, producing multiple species as indicated by ^{1}H NMR spectroscopy of the reaction mixtures.

B.1.2.2 Characterization and Reactivity Studies of Molybdenum(V) and Tungsten(V) Complexes

Starting from the mono(oxo) molybdenum(VI) precursor MoOCl$_4$, selective complex formation was only observed using the deprotonated pro-ligands (MOC$_6$H$_2$-R^1-4-R^2-6)$_2${S-R^3-S} (R^1 = R^2 = ^{t}Bu, R^3 = (CH$_2$)$_2$ for **2a**; R^1 = Me, R^2 = ^{t}Bu, R^3 = (CH$_2$)$_2$ for **2b**; R^1 = R^2 = ^{t}Bu, R^3 = cyclohexanediyl for **2c**; M = Li, Na, K) in toluene (Scheme B.1.6). Using the pro-ligands [(OSSO)H$_2$] or other solvents such as THF or benzene resulted in the formation of blue reaction mixtures, indicating the formation of unidentified oligomeric bridging Mo-oxo complexes.[26] Monitoring the reaction of [(OSSO)Li$_2$] with MoOCl$_4$ in toluene-d_8 by ^{1}H NMR spectroscopy revealed the formation of a C_2-symmetric intermediate **A**, which was subsequently consumed in favor of the paramagnetic complex[(OSSO)MoOCl] (**2a**). Analyzing the reaction mixture by GC MS did not reveal the formation of chlorinated toluene derivatives, excluding toluene as reducing agent. Presumably, the intermediate **A** disproportionates to insoluble oligomeric species which may account for the low yield of 37%. Consequently, the already reduced molybdenum(V) precursor MoOCl$_3$ was used to prepare the [(OSSO)MoOCl] complexes which gave the respective complexes in much higher yields of 73 – 89% (Scheme B.1.6).[27] The [(OSSO)MoOCl] complexes were obtained as moisture- and air-sensitive purple solids. The EI mass spectra show prominent peaks at m/z = 649.1486 (**2a**), 565.053 (**2b**) and 702.7 (**2c**) which can be assigned to [(OSSO)MoOCl]$^+$. A satisfactory elemental composition could not be determined by combustion analysis, due to presumable carbide formation and contamination by metal halide salts.[28]

Results and Discussion

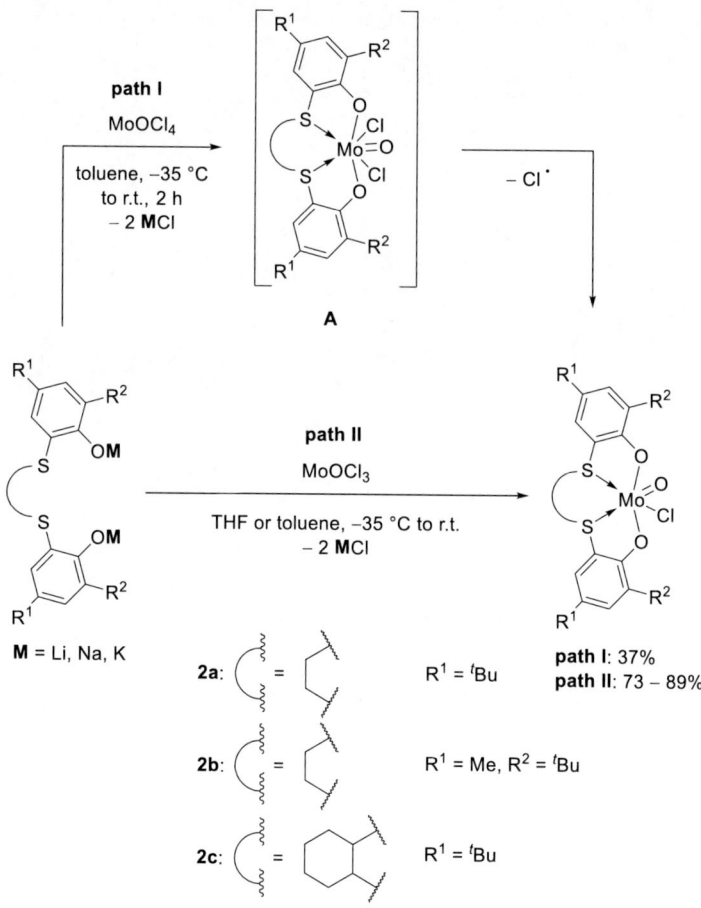

Scheme B.1.6. Synthesis of the [(OSSO)MoOCl] complexes (**2a**, **2b**, **2c**) starting from MoOCl$_4$ via intermediate **A** in 37% yield and starting from MoOCl$_3$ in 73 – 89% yield.

The analogous tungsten complex [(OSSO)WOCl] was prepared using the metal precursor WOCl$_4$ (Scheme B.1.7). Due to the lower reactivity of the WOCl$_4$ precursor compared to the Mo analogue, the potassium salt of the pro-ligand [(OSSO)K$_2$] was used to obtain [(OSSO)WOCl] in isolable amounts. Starting from the mono-oxo tungsten(VI) precursor WOCl$_4$, selective complex formation was observed using the deprotonated pro-ligand (KOC$_6$H$_2$-R^1-4-R^2-6)$_2${S-R^3-S} (R^1 = R^2 = ^{t}Bu, R^3 = cyclohexanediyl for **4**) in toluene (Scheme B.1.7). Complex **4** was obtained as a moisture- and air-sensitive purple/brown solid in 52% yield. The EI mass spectrum shows a prominent peak at m/z = 788.1 which can be assigned to [(OSSO)WOCl]$^+$.

Similar to the Mo analogues, a satisfactory elemental composition could not be determined by combustion analysis.

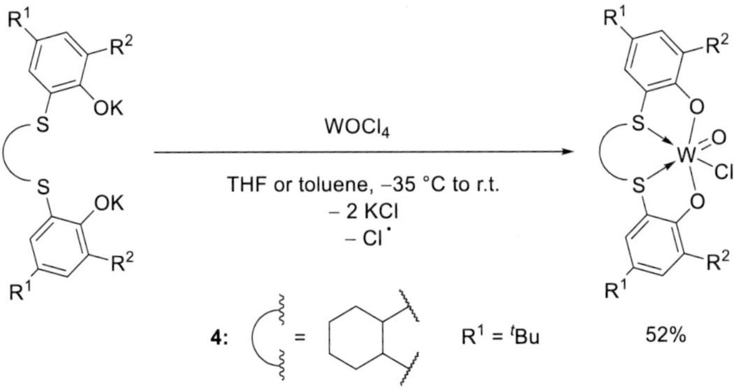

Scheme B.1.7. Synthesis of the [(OSSO)WOCl] complex (**4**) starting from WOCl₄ in 52% yield.

Single crystals of **2b** were grown from toluene/*n*-hexane and the molecular structure was established by X-ray diffraction analysis (Figure B.1.6). Similar to **1b**, the metal center is coordinated in distorted octahedral geometry with the chloro- and oxo-ligands coordinated in the equatorial plane and the (OSSO)-ligand in helical *cis-α* coordination.[17, 20] The Mo=O bond length (1.677(3) Å) is in a typical range for other Mo(V) mono-oxo complexes.[29] Due to the *trans* influence of the oxo-ligand, the Mo1–S2 distance of 2.844(13) Å is longer than Mo1–S1 (2.5144(12) Å) and those in the analogous dioxo molybdenum(VI) complex (2.7348(13) Å and 2.6955(13) Å).[16] **2b** and **1b** have similar Mo–Cl (2.3259(12) Å for **2b** and 2.3306(10) Å for **1b**) and Mo–S (2.5144(12) Å for **2b** and 2.5298(9) Å for **1b**) distances *trans* to the chloro-ligand.[20]

Results and Discussion

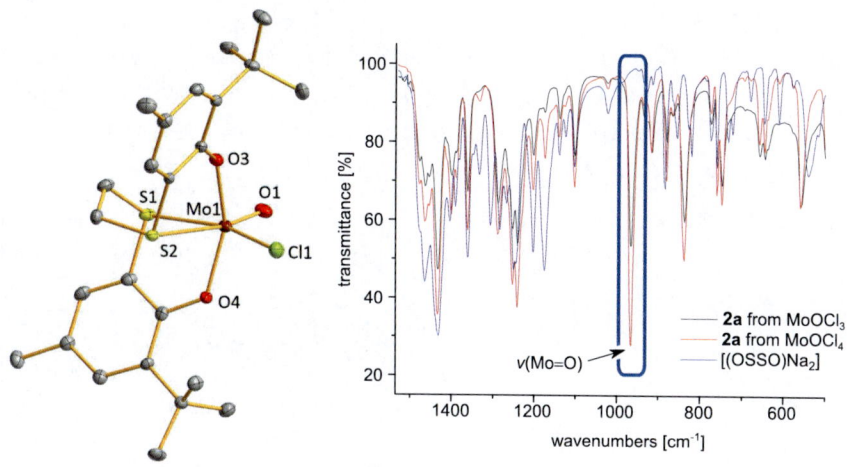

Figure B.1.6. Molecular structures of **2b** with 50% displacement ellipsoids (left); all hydrogen atoms are omitted for clarity. Selected bond lengths (Å) and angles (°): Mo1–O1 1.677(3), Mo1–O2 1.973(3), Mo1–O3 1.964(3), Mo1–S1 2.5144(12), Mo1–S2 2.8448(13), Mo1–Cl12.3259(12), O1–Mo1–Cl1 104.87(12), O2–Mo1–O3 151.75(13).[20] Excerpt of the IR spectrum of [(OSSO)Na$_2$] and **2a**, synthesized through path I or II (right).

The IR spectra of the [(OSSO)MOCl] (M = Mo, W) complexes show absorptions at $\tilde{\nu}$ = 965 cm^{-1} (**2a**), 967 cm^{-1} (**2b**), 957 cm^{-1} (**2c**) and 968 cm^{-1} (**4**) which are in a typical region for terminal M=O (M = Mo, W) stretching vibrations.[11] These vibrational modes were not observed in the IR spectra of the deprotonated [(OSSO)M$_2$] (M = Li, Na, K) pro-ligands and are shifted compared to the metal precursors MoOCl$_n$ (n = 3, 4) (Figure B.1.6). Compounds **2a**, **2b**, **2c** and **4** are paramagnetic and do not produce any resonances in their respective ^{1}H NMR spectra. **2a** and **2b** exhibit magnetic moments of μ_{eff} = 1.5 μ_B for **2a** and μ_{eff} = 1.8 μ_B for **2b** in solution (DCM-d_2), which were determined by Evans Method.[21, 22] The magnetic moments are close to the spin-only magnetic moment μ_s = 1.7 μ_B for a d^1 complexes with S_{total} = 1/2.[23]

EPR spectroscopy was used to gain further insight into the chemical environment of the d^1 complexes **2a** and **2b**. Figure B.1.7 shows the recorded and simulated EPR spectra of **2a** and **2b** in THF at 298 K. Coupling of the electron spin with the molybdenum center (I = 0 (75% abundance), 5/2 (25% abundance)) produces isotropic signals with 1 + 6 lines (2I + 1; for I = 0 one singlet with 75% intensity and for I = 5/2 one sextet with each 4% intensity) with hyperfine coupling constants A_{iso} of 130 MHz and g_{iso} values of 1.94.[30] These values are similar to the analogous [(SSSS)MoOCl] complex with a g_{eff} value of 1.974 and a hyperfine coupling constant A of 3.9 mT (approximately 108 MHz).[11] Superhyperfine interaction was not observed. The

Group 6 Metal Complexes Featuring a Tetradentate (OSSO)-Type Ligand

signal remains isotropic in frozen solutions at 77 K. Simulation of the experimental EPR spectra confirms the assignment of the resonances to a molybdenum d^1 center. The results of the EPR spectra and the magnetic properties of **2a** and **2b** consistently indicate one localized electron at the metal center with S_{total} = 1/2 and molybdenum in the formal oxidation sate +V. Complex **4** could not be characterized by EPR spectroscopy since only a very weak and broad signal was observed using an X-band spectrometer.

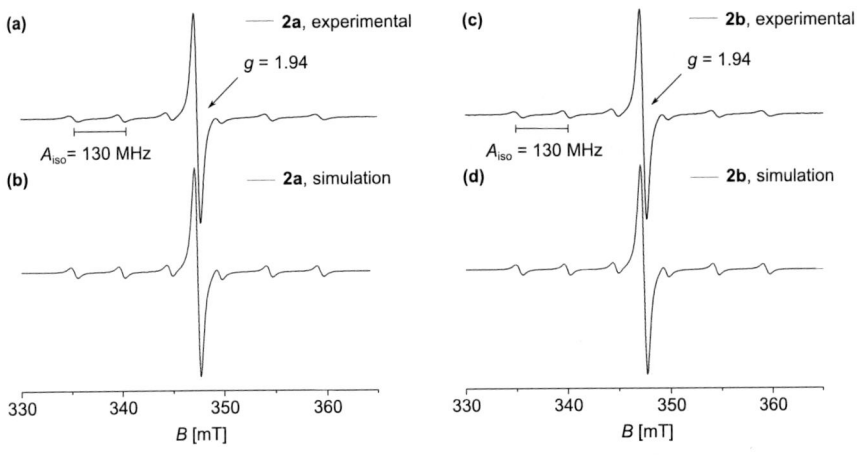

Figure B.1.7. Experimental X-band EPR spectra of (a) **2a** and (c) **2b** in THF at 298 K and simulated EPR spectra of (b) **2a** and (d) **2b** at 298 K.[20]

To determine whether a chemical reduction of the [(OSSO)MoOCl] complexes may facilitate an access to a mono-oxo molybdenum(IV) complexes, the compounds were characterized by cyclic voltammetry (Figure B.1.8(a), (b)). The electrochemical study of **2a** and **2b** in DCM revealed one reversible oxidation at $E_{1/2}(1)$ = 0.57 V for **2a** and at $E_{1/2}(1)$ = 0.48 V for **2b**, as well as a quasi-reversible reduction at $E_{1/2}(2)$ = –1.21 V for **2a** and at $E_{1/2}(2)$ = –1.25 V for **2b** (vs Fc/Fc⁺, compound (2 mM), electrolyte [^{n}Bu₄N][PF₆] (100 mM), scan rate 100 mV s⁻¹). The nature of the redox events of **2b** was further elucidated through recording cyclic voltammograms at different scan rates and plotting the ratio of the peak currents against $v^{1/2}$ (Figure B.1.8(c)). Compared to **1a** (Figure B.1.5), a larger deviation of the peak currents of the oxidative event at $E_{1/2}(1)$ = 0.48 V from linearity is observed (Figure B.1.8(c)). Although the scattering of the measured ratios is larger, a linear trend is suggested, which indicates the presence of a quasi-reversible to reversible electron transfer process. On the other hand, the ratios of the peak currents of the reductive event at $E_{1/2}(2)$ = –1.25 V are not located on a horizontal line, thus indicating a quasi-reversible to irreversible electron transfer process. The two redox events correspond to the one-electron-oxidation and one-electron-reduction of the

Mo(V) complexes. The reduction potential of **2a** and **2b** is lower than that of the analogous [(SSSS)MoOCl] complex (−0.18 V in DMF vs. SCE, corresponding to approx. −0.63 V vs. Fc/Fc⁺).[11, 31] Similar to **1a**, the quasi-reversible reduction may be attributed to dissociation of one chloro ligand and subsequent reactions. Complex **4** produces a complex cyclic voltammogram with multiple redox events which are not undoubtedly assignable. Similar electrochemical differences between analogous molybdenum and tungsten complexes have been reported for dioxo tungsten(VI) and molybdenum(VI) complexes featuring salan (ONNO)-type ligands.[32]

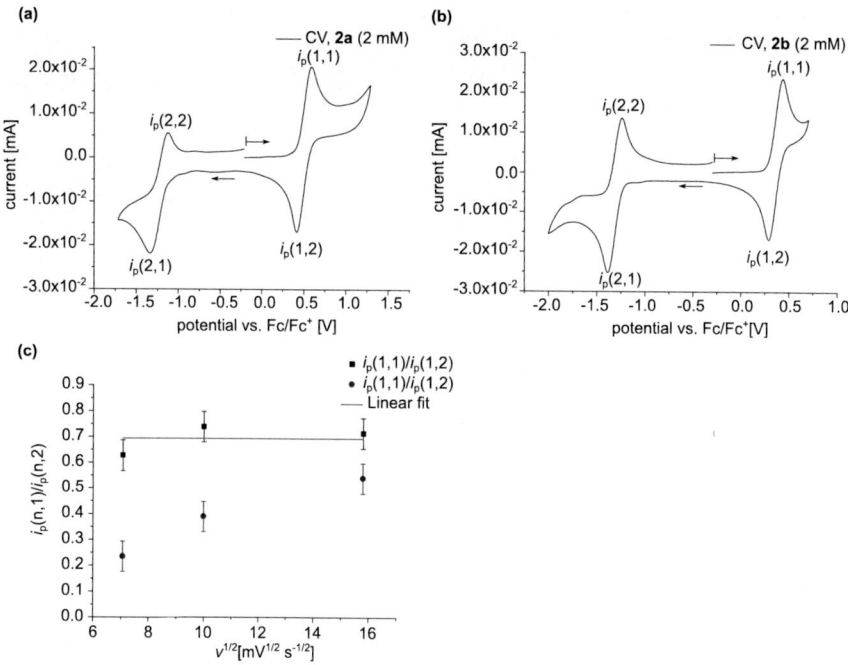

Figure B.1.8. Cyclic voltammograms of (a) **2a** (2 mM) and of (b) **2b** (2 mM) in DCM at 298 K, scan rate 100 mV s^{-1} (**2a**) and 250 mV s^{-1} (**2b**), electrolyte [^{n}Bu$_4$N][PF$_6$] (100 mM) referenced to the Fc/Fc⁺ couple; (c) plot of the ratios of the peak currents (i_p(n,1)/i_p(n,2)) of **2b** against $v^{1/2}$ (v = scan rate).[20]

The electrochemical characterization of **2a** and **2b** indicated that oxidation may selectively produce a diamagnetic Mo(VI) complexes. However, attempts to selectively oxidize **2a** with chemical oxidants such as [Ag][B(C$_6$F$_5$)$_4$] or [N(C$_6$H$_4$Br)$_3$][B(C$_6$F$_5$)$_4$] produced multiple species in the respective ¹H NMR spectra due to presumable abstraction of the chloro ligand (Scheme B.1.8). Different reducing agents were tested to access a[(OSSO)MoO] complex. Treating **2a** with KC$_8$ did not result in any reaction. Following the reduction of **2a** with cobaltocene by ¹H NMR spectroscopy in THF-d_8 indicated complete consumption of cobaltocene. However, the

Group 6 Metal Complexes Featuring a Tetradentate (OSSO)-Type Ligand

product could not be isolated due to decomposition and formation of an insoluble precipitate. Reduction of **2a** with silyl reducing agents were unselective due to abstraction of the oxo ligand to produce $(Me_3Si)_2O$ and abstraction of the chloro ligand to produce Me_3SiCl as followed by ^1H NMR spectroscopy of the THF-d_8 solutions.[33, 34]

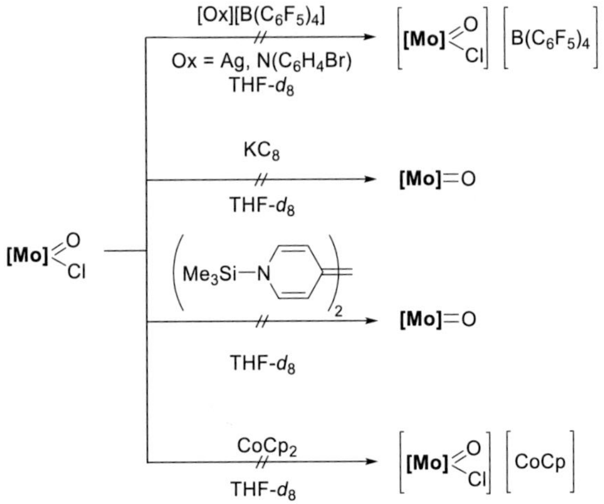

Scheme B.1.8. Reactivity studies of **2a** ([**Mo**] = (OSSO)Mo).

B.1.2.3 Reactivity Studies of Molybdenum(VI) Complexes

Previous studies indicated that the dioxo complex [(OSSO)MoO$_2$] (**3**) catalyzes the OAT reaction of DMSO with PPh$_3$.[17] Since the OAT reaction starting from **3** involves a [(OSSO)MoO] intermediate, the reduction of the high oxidation state dioxo complex **3** with phosphines through abstraction of one oxo-ligand was investigated. Complex **3** was prepared according to literature conditions through protonolysis of MoO$_2$Cl$_2$ with [(OSSO)H$_2$].[16] An electrochemical study of complex **3** in DCM was performed, revealing a quasi-reversible reduction at $E_{1/2}$ = −1.81 V which corresponds to the one-electron reduction of Mo(VI) to Mo(V) (vs. Fc/Fc$^+$, compound (2 mM), electrolyte [^{n}Bu$_4$N][PF$_6$] (100 mM), Figure B.1.9). Similar to the dichloro molybdenum(IV) and mono-oxo molybdenum(V) complexes featuring the (OSSO)-type ligand, the reduced species further reacts upon reduction resulting in additional redox events in the cyclic voltammograms. The [(OSSO)MoOCl] and [(OSSO)MoO$_2$] complexes exhibit reductions at more negative potentials and oxidations at more positive potentials compared to the (SSSS)-analogues.[11, 31] The difference of the redox properties is caused by a change of the Lewis-basic properties of the ligand backbone through exchanging the soft thiophenolato ligand against the hard phenolato one. Similar shifts of the redox potentials have been observed in other transition metal complexes through exchanging the ligand donor-atom to the higher homologue.[35] For all molybdenum complexes featuring (OSSO)-type ligands, redox events with smaller peak currents were observed at more positive or negative potentials which may be attributed to ligand centered electrochemical process, as previously observed for salen-type complexes.[36]

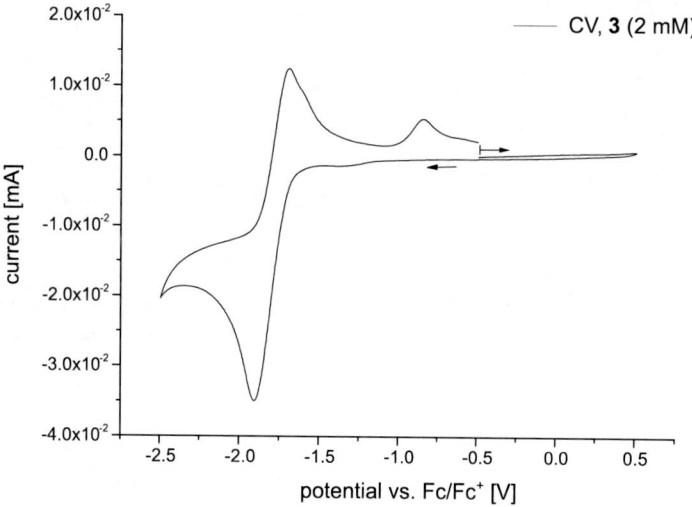

Figure B.1.9. Cyclic voltammograms of **3** (2 mM) in DCM at 298 K, scan rate 100 mV s^{-1}, electrolyte [^{n}Bu$_4$N][PF$_6$] (100 mM) referenced to the Fc/Fc$^+$ couple.[20]

Monitoring the reduction of **3** with PPh$_3$, PMePh$_2$ or PMe$_3$ in THF-d_8 at 60 °C with ^{1}H and ^{31}P{^{1}H} NMR spectroscopy, revealed the formation of multiple phosphorous containing species after several days. The formation of multiple species may be rationalized by dimerization of the reduced complex which has also been reported for the [(SSSS)MoO$_2$] analogue following reduction with 0.5 equiv. PPh$_3$.[11] Addition of 1 equiv. of B(C$_6$F$_5$)$_3$ to **3** in THF-d_8 produces the [(OSSO)MoO$_2$][B(C$_6$F$_5$)$_3$] (**[3][B(C$_6$F$_5$)$_3$]**) adduct as indicated by a shift of the signal in the ^{11}B{^{1}H} NMR spectrum from δ 57.2 ppm to δ 2.6 ppm. Upon treatment of **[3][B(C$_6$F$_5$)$_3$]** with PPh$_3$ at 60 °C, the Lewis acid B(C$_6$F$_5$)$_3$ is abstracted to afford the starting complex **3** and [PPh$_3$][B(C$_6$F$_5$)$_3$] (Scheme B.1.9) as indicated by the shift of the singlet resonance from δ 5.4 ppm to δ 23.7 ppm in the ^{31}P{^{1}H} NMR spectrum and by the shift of the singlet resonance from δ 2.6 ppm to δ 2.9 ppm in the ^{11}B{^{1}H} NMR spectrum. ^{1}H NMR spectroscopy also indicated the formation of the starting complex as indicated by the two resonances exhibited by the ^{t}Bu substituents at δ 1.42 ppm and δ 1.30 ppm. Hence, stabilization of the [(OSSO)MoO$_2$] complex through a Lewis acid-base-adduct was not successful.

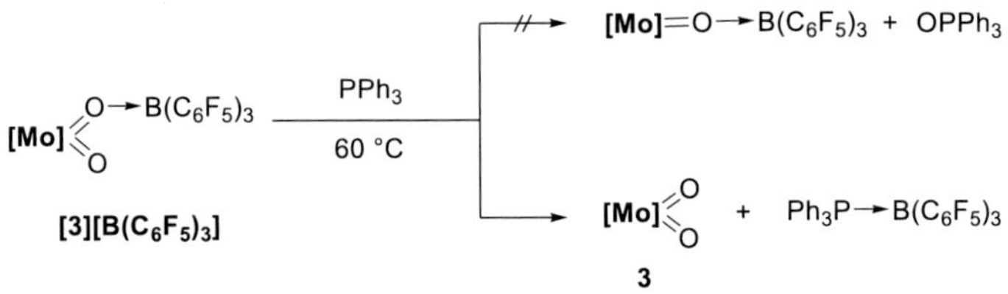

Group 6 Metal Complexes Featuring a Tetradentate (OSSO)-Type Ligand

Scheme B.1.9. Formation of the Lewis acid/base adduct [3][B(C$_6$F$_5$)$_3$] and regeneration of the starting complex 3 upon treatment with PPh$_3$.

B.1.2.4 Catalytic Oxygen Atom Transfer (OAT) Reactions

The series of molybdenum complexes in formal oxidation states of IV, V and VI enabled investigations on the influence of the initial oxidation state onto the catalytic OAT reactions. The catalytic activity of **1a**, **1b**, **2a**, **2b** and **3** for the reduction of DMSO with PPh$_3$ under neat conditions in DMSO-d_6 was tested. Complexes **1a – 2b** showed high catalytic activity with complete conversion of PPh$_3$ to OPPh$_3$ indicated by complete consumption of the signal at δ −6.8 ppm in favor of δ 25.5 ppm in the ^{31}P{^{1}H} NMR spectra (Table B.1.1). **3** exhibited low catalytic performance with 4% conversion of PPh$_3$ to OPPh$_3$, determined through integration of the aromatic resonances at δ 7.66 − 7.50 ppm for OPPh$_3$ and at δ 7.44 − 7.19 ppm for PPh$_3$ in the ^{1}H NMR spectrum. The low catalytic activity of **3** is rationalized by dimerization of the active species as indicated by the unselective reduction of **3** upon treatment with phosphines. Similar reactivity was also reported for the [(SSSS)MoO$_2$] analogue after reduction with 0.5 equiv. of PPh$_3$.[11] Since the molybdenum(V) complexes **2a** and **2b** showed high activity, the formation of a catalytically inactive molybdenum(V) complex upon reduction of **3** was ruled out.

To gain further insight into the catalytic OAT reactions, the molybdenum(IV) and (V) catalysts were tested under similar conditions in THF-d_8 or THF/THF-d_8 solutions. ^{31}P{^{1}H} NMR spectroscopy of the solutions revealed complete conversion of PPh$_3$ to OPPh$_3$ after 2 h in case **1a** and **1b** were used as catalysts, indicated by complete consumption of the signal at δ −7.3 ppm for PPh$_3$ in favor of δ 21.5 ppm for OPPh$_3$. The conversion was determined through integration of the aromatic resonances at δ 7.73 − 7.40 ppm for OPPh$_3$ and at δ 7.35 − 7.24 ppm for PPh$_3$ in the ^{1}H NMR spectrum. In case **2a** is used as a catalyst, a lower activity with only 10% conversion of PPh$_3$ to OPPh$_3$ was observed after heating the solution for 66 h at 60 °C. The higher activity of the [(OSSO)MoCl$_2$] complexes **1a** and **1b** is explained by their high oxophilicity, which facilitates fast oxidation by DMSO. During the OAT catalytic cycle, the molybdenum(V) complex **2a** needs to form a Mo(V)/Mo(III) couple which may account for the decreased catalytic performance compared to the Mo(VI)/Mo(IV) couple. Similar to performing the reaction under neat conditions, **3** converted less than 1% of PPh$_3$ to OPPh$_3$ after 2 h.

Table B.1.1. Results of the OAT reaction from DMSO to PPh$_3$. The conversions were determined by NMR spectroscopy.

entry	catalyst	catalyst loading [mol%]	solvent	time [h]	temperature [°C]	conversion [%]
1[a]	1a	5	DMSO-d_6	20	80	>99
2[a]	1b	5	DMSO-d_6	20	80	>99
3[a]	2a	5	DMSO-d_6	20	80	>99
4[a]	2b	5	DMSO-d_6	20	80	>99
5[a]	3	5	DMSO-d_6	20	80	4
6[b]	1a	5	THF-d_8	66	60	>99
7[b]	2a	5	THF-d_8	66	60	10
8[c]	1a	5	THF/THF-d_8	2	60	>99
9[c]	1b	5	THF/THF-d_8	2	60	>99
10[c]	3	5	THF/THF-d_8	2	60	<1
11[c]	[MoCl$_4$(NCMe)$_2$]	5	THF/THF-d_8	2	60	>99
12[c]	MoOCl$_3$	5	THF/THF-d_8	2	60	6
13[d]	1a	1	THF/THF-d_8	0.5	60	99
14[d]	1a	1	THF/THF-d_8	0.5	25	7
15[d]	[MoCl$_4$(NCMe)$_2$]	1	THF/THF-d_8	0.5	60	>99
16[d]	[MoCl$_4$(NCMe)$_2$]	1	THF/THF-d_8	0.08	25	>99

[a] catalyst (5.3 μmol), DMSO-d_6 (0.6 mL), PPh$_3$ (266 μmol); [b] catalyst (5.3 μmol), PPh$_3$ (106 μmol), DMSO (106 μmol); [c] catalyst (5.3 μmol), PPh$_3$ (106 μmol), DMSO (106 μmol, 3.5 M in THF); [d] catalyst (6.0 μmol), PPh$_3$ (600 μmol), DMSO (600 μmol, 3.5 M in THF).

To determine the reason for the high catalytic activity of **1**, complex **1a** was pre-oxidized with DMSO in THF/THF-d_8 for 2 h at 60 °C. Similar to **2a** or **2b** the solution turned dark purple, indicating presumable formation of the less active [(OSSO)MoOCl] complex. After adding the oxygen atom acceptor PPh$_3$ and continued heating for 2 h at 60 °C, only 7% of PPh$_3$ were converted to OPPh$_3$, which is similar to the activity of **2a**. These observations are explained by the proposed catalytic cycle (Scheme B.1.10) which proceeds *via* a mono-oxo molybdenum(VI) intermediate, similar to the one observed for natural DMSO reductase.[2, 37-39] In the first step, the dichloro complex is oxidized by DMSO *via* the oxo-bridged transition state **B** to subsequently generate the molybdenum(VI) intermediate **A** and release dimethyl sulfide. The intermediate species **A** is unstable and reacts with PPh$_3$ *via* the oxo-bridged transition state **C** to regenerate the initial dichloro complex and OPPh$_3$, closing the catalytic cycle. In absence of PPh$_3$ as an oxygen atom scavenger, **A** decomposes to the molybdenum(V) complex **2** as was observed during the preparation of **2a** starting from MoOCl$_4$. The high catalytic activity of **1a** and **1b** is caused by the formation of the highly reactive molybdenum(VI) intermediate **A**. In contrast to the unstable species **A**, **3** is relatively stable in the +VI oxidation state, as reported for the [(SSSS)MoO$_2$] analogue.[11]

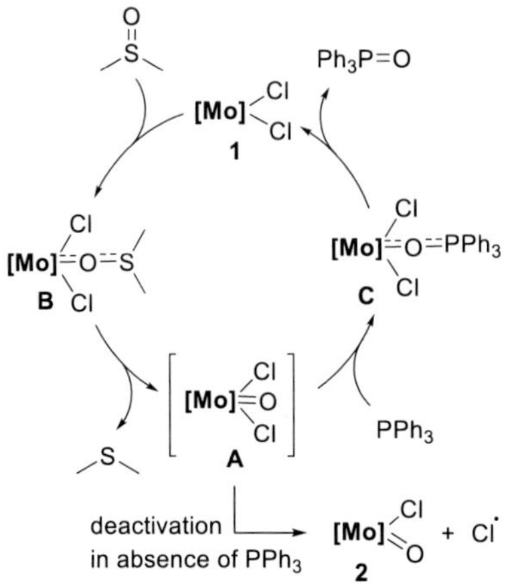

Group 6 Metal Complexes Featuring a Tetradentate (OSSO)-Type Ligand

Scheme B.1.10. Proposed catalytic cycle of the reduction of DMSO with PPh$_3$ as catalyzed by **1** and decomposition in absence of an oxygen atom scavenger (**[Mo]** = (OSSO)Mo).[20]

Due to the decomposition of **1** upon oxidation with DMSO in the absence of an oxygen atom scavenger, **1** exhibits lower catalytic activity after prior oxidation. The low stability of the intermediate **A** is comparable to the stability of the metal precursor MoOCl$_4$, which is thermally unstable and slowly decomposes to MoOCl$_3$.[40] The catalytic activity of the starting metal precursor MoOCl$_3$ and [MoCl$_4$(NCMe)$_2$] was also tested. Similar to **2**, the molybdenum(V) precursor exhibited a low catalytic activity and converted only 6% of PPh$_3$. Upon addition of PPh$_3$ to [MoCl$_4$(NCMe)$_2$] the complex immediately dissolved in THF-d_8 to produce an orange solution. Most likely the bis(phosphine) complex [MoCl$_4$(PPh$_3$)$_2$] is generated *in situ*, making the catalyst soluble in THF. The proposed complex [MoCl$_4$(PPh$_3$)$_2$] showed remarkable high activity and completely oxidized PPh$_3$ in less than 5 min at room temperature, whereas **1a** only showed a conversion of 7% in 30 min under the same reaction conditions. Due to the high oxophilicity of the catalyst, addition of DMSO to a suspension of [MoCl$_4$(NCMe)$_2$] in THF-d_8 leads to complete deactivation of the catalyst. To ensure high catalytic activity, the [MoCl$_4$(PPh$_3$)$_2$] complex needs to be generated *in situ* first before DMSO is added to the reaction mixture. At a catalyst loading of 1 mol%, the reaction was already completed once the NMR measurement had been finished (< 5 min), corresponding to a turnover frequency of more than 1590 h^{-1}. The catalytic activity of the [MoCl$_4$(NCMe)$_2$] complex surpasses the one reported for Mo(VI) complexes such as [MO$_2$Cl$_2$(dmf)$_2$] or [BMIm]$_2$[MoO$_2$(NCS)$_4$].[41] Similar to **1a**, these complexes require elevated temperatures to ensure complete deoxygenation of sulfoxides.

B.1.3 Summary and Outlook

The magnetic and electrochemical properties of the dichloro molybdenum(IV) complexes **1a** and **1b** were determined and their reactivity was further elucidated. Contrary to previous studies, the analogous tungsten complex could not be prepared on a preparative scale.[17] The complexes exhibit a d^2-spin system and are paramagnetic. Electrochemical characterization revealed one reversible oxidation and one quasi-reversible reduction event. Attempts to chemically oxidize or reduce the complexes were unsuccessful. Ligand exchange reactions to produce dialkyl, dihydrido or mono-oxo molybdenum(IV) complexes did not afford the desired compounds.

The synthesis of the mono-oxo molybdenum(V) and -tungsten(V) complexes **2a**, **2b**, **2c** and **4** stabilized by (OSSO)-type ligands was reported (Scheme B.1.11). The molybdenum complexes were prepared through salt metathesis either starting from the molybdenum(VI) precursor $MoOCl_4$ or the molybdenum(V) precursor $MoOCl_3$ with the deprotonated pro-ligands [(OSSO)M$_2$] (M = Li, Na, K). For the preparation of the analogous tungsten complex, the potassium salt of the pro-ligand [(OSSO)K$_2$] was used due to the lower activity of the $WOCl_4$ metal precursor. Complexes **2a** and **2b** were spectroscopically and electrochemically characterized, and their structure was confirmed by X-ray diffraction on single crystals of **2a**. **2a** and **2b** are paramagnetic and exhibit a d^1 spin-system as determined by EPR spectroscopy and Evans method. Electrochemical characterization indicated one reversible oxidation and one quasi-reversible reduction event.

The electrochemical properties of the dichloro molybdenum(IV) and the mono-oxo molybdenum(V) complexes were compared with the dioxo molybdenum(VI) complex **3**.[16] Electrochemical characterization of **3** showed one quasi-reversible reduction event. Attempts to chemically reduce **3** with phosphines to afford mono-oxo complexes produced oligomeric species.

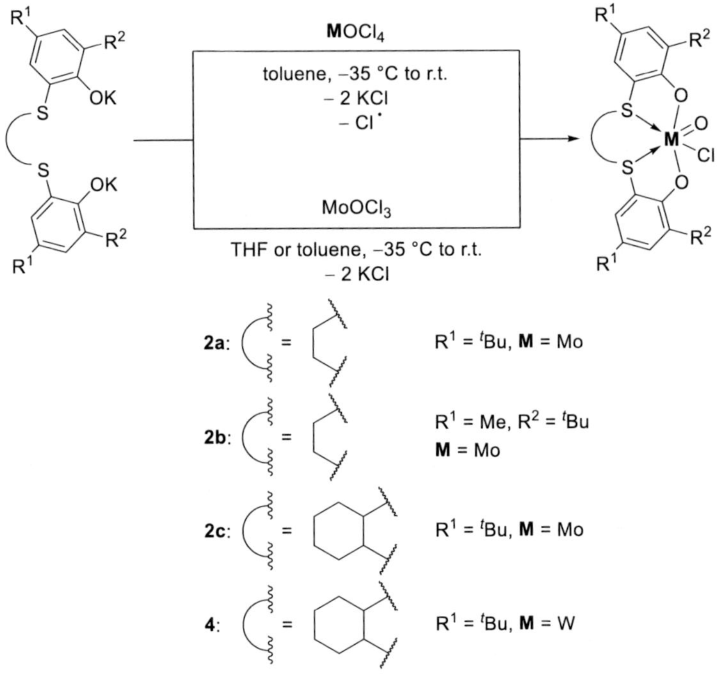

Group 6 Metal Complexes Featuring a Tetradentate (OSSO)-Type Ligand

Scheme B.1.11. Synthesis of mono-oxo molybdenum(V) and tungsten(V) complexes featuring (OSSO)-type ligands.

The catalytic performance of the molybdenum complexes in the initial oxidation states +IV, +V and +VI in OAT reactions was studied. All molybdenum complexes **1 – 3** catalyze the model reaction of the DMSO reductase family, in which PPh_3 is oxidized by DMSO. The lowest activity was observed for **3** which is explained by the high relative stability of the +VI oxidation state. The highest activity was found for the dichloro molybdenum(IV) complexes **1a** and **1b**. The high catalytic activity of **1** is rationalized by the formation of the highly reactive intermediate **A**. The latter decomposes to **2** if no oxygen atom scavenger is present. Surprisingly the tetrachloro complex [$MoCl_4(MeCN)_2$] surpasses the catalytic activity of **1**.

Further research should focus on the highly active [$MoCl_4(NCMe)_2$] catalyst, which requires fewer synthetic steps compared to the complexes stabilized by tetradentate (OSSO)-type ligands. The simple preparation and high catalytic activity make this catalyst interesting for organic synthesis. To further elucidate the catalytic properties, substrate featuring different functional groups should be explored, broadening the substrate scope.

Results and Discussion

B.1.4 Experimental

B.1.4.1 General Considerations

[(OSSO)H$_2$] pro-ligands,[42-45] MoOCl$_3$,[27] [MoCl$_4$(NCMe)$_2$],[19] and [(OSSO)MoO$_2$] (3)[16] were prepared according to literature procedures. All other chemicals were purchased form commercial sources and used as received without further purification.

B.1.4.2 Synthesis of the Deprotonated Pro-Ligands

Typical procedure to prepare [(OSSO)Li$_2$]

^{n}BuLi (2.20 equiv, 2.5 M in n-hexane) was added dropwise to a stirred suspension of the pro-ligand [(OSSO)H$_2$] (1.00 equiv.) in Et$_2$O while cooling with a water/ice bath. The solution was stirred for 2 h at 25 °C. All volatiles were removed under reduced pressure to produce a colorless solid. The solid was rinsed with n-hexane and dried under reduced pressure to afford [(OSSO)Li$_2$] as a colorless solid.

Typical procedure to prepare [(OSSO)Na$_2$]

A solution of the pro-ligand [(OSSO)H$_2$] (1.00 equiv.) in THF was added dropwise to a stirred suspension of NaH (2.00 equiv) in THF while cooling with a water/ice bath. The suspension was stirred for 17 h at 25 °C until all gas evolution subsided. All volatiles were removed under reduced pressure to afford [(OSSO)K$_2$] as colorless solid.

Typical procedure to prepare [(OSSO)K$_2$]

A solution of the pro-ligand [(OSSO)H$_2$] (1.00 equiv.) in THF was added dropwise to a stirred suspension of KH (2.00 equiv) in THF while cooling with a water/ice bath. The suspension was stirred for 17 h at 25 °C until all gas evolution subsided. All volatiles were removed under reduced pressure to afford [(OSSO)K$_2$] as colorless solid.

B.1.4.3 Synthesis of Molybdenum Complexes

Dichloro{1,2-dithiabutanediyl-2,2'-bis(4,6-di-tert-butyl-phenolato)}molybdenum(IV) (1a)

A solution of the [(OSSO)H$_2$] pro-ligand 1,2-dithiabutanediyl-2,2'-bis(4,6-di-*tert*-butyl-phenol) (1.53 g, 3.05 mmol, 1.00 equiv.) and [MoCl$_4$(NCMe)$_2$] (974 mg, 3.05 mmol, 1.00 equiv.) in THF (100 mL) was stirred for 18 h at 60 °C. All volatiles were removed under reduced pressure. The orange/red solid was rinsed with n-pentane (10 mL) and, subsequently, extracted with THF. Crystallization from THF/n-hexane afforded **1a** as red crystals (904 mg, 1.35 mmol, 44%). The compound is paramagnetic and produces signals over a large range of chemical shifts. ^{1}H **NMR** (DCM-d_2): δ 19.35 (s, 2H), 2.01 (br s, 18H, C(CH$_3$)$_3$), 1.73 (s, 2H), 1.45 (br s, 18H, C(CH$_3$)$_3$), −9.45 (s, 2H), −85.91 (s, 2H). **Elemental Analysis**: Calcd. for C$_{30}$H$_{44}$Cl$_2$O$_2$S$_2$Mo: C 53.97, H 6.64; found: C 53.96, H 6.74. **CV** (in DCM, **1a** (2 mM), [^{n}Bu$_4$N][PF$_6$] (100 mM),

Group 6 Metal Complexes Featuring a Tetradentate (OSSO)-Type Ligand

100 mV s^{-1}): $E_{1/2}$ (1) = 0.50 V (reversible), $E_{1/2}$ (2) = −0.98 V (quasi-reversible). **Magnetic Properties**: μ_{eff} = 2.3 μ_B (Evans method, 298 K, DCM).

rac-Dichloro{dithiocyclohexanediyl-2,2'-bis(4,6-dicumylphenolato)}molybdenum(IV) (1b)

The synthesis was performed analogously to that for **1a**, according to previously reported procedures.[17] The respective pro-ligand [(OSSO)H$_2$] (100 mg, 124 µmol, 1.00 equiv.) and [MoCl$_4$(NCMe)$_2$] (39.7 mg, 124 µmol, 1.00 equiv.) were used. **1b** was purified through rinsing with n-pentane and extraction with DCM. Removal of all volatiles under reduced pressure afforded **1b** as an orange/red powder (114 mg, 118 µmol, 95%). ^{1}H NMR (DCM-d$_2$): δ 19.74 (br s, 2H), 11.34 (br s, 2H), 7.83 (br s, 5H), 7.69 (br s, 5H), 7.27 (t, $^3J_{HH}$ = 7.3 Hz, 4H, H_{aryl}), 7.11 (d, $^3J_{HH}$ = 7.7 Hz, 4H, H_{aryl}), 7.04 (t, $^3J_{HH}$ = 7.3 Hz, 2H, H_{aryl}), 3.97 (br s, 2H), 2.59 (br s, 6H, CH$_3$), 1.96 (s, 6H, CH$_3$), 1.86 (s, 6H, CH$_3$), 0.97 (s, 2H), 0.08 (br s, 2H), −0.69 (br s, 2H), −7.99 (br s, 2H). **Elemental Analysis**: Calcd. for C$_{54}$H$_{58}$Cl$_2$O$_2$S$_2$Mo: C 66.86, H 6.03; found: C 66.37, H 6.16. **CV** (in DCM, **1b** (2 mM), [^{n}Bu$_4$N][PF$_6$] (100 mM), 100 mV s^{-1}): $E_{1/2}$ (1) = 0.50 V (reversible), $E_{1/2}$ (2) = −1.12 V (quasi-reversible). **Magnetic Properties**: μ_{eff} = 2.6 μ_B (Evans method, 298 K, DCM).

Chloro(oxo){1,2-dithiabutanediyl-2,2'-bis(4,6-di-tert-butyl-phenolato)}molybdenum(V) (2a)

Method A: A suspension of the deprotonated pro-ligand [(OSSO)K$_2$] (100 mg, 173 µmol, 1.00 equiv.) in toluene (5 mL) was added to a green solution of MoOCl$_4$ (43.8 mg, 173 µmol, 1.00 equiv.) in toluene (5 mL) at −35 °C. The reaction mixture was stirred for 2 h at 25 °C to produce a red/purple suspension. The solid was separated through centrifugation and the solution was collected. The solid was rinsed with toluene and extracted. The combined solutions were collected, and all volatiles were removed under reduced pressure. The red/purple solid was rinsed with n-pentane and extracted with DCM. Drying under reduced pressure yielded **2a** as a purple solid (41.5 mg, 64.0 µmol, 37%).

Method B: A suspension of the deprotonated pro-ligand [(OSSO)Li$_2$] (623 mg, 972 µmol, 1.00 equiv.) in toluene (10 mL) was added dropwise to a brown suspension of MoOCl$_3$ (212 mg, 972 µmol, 1.00 equiv.) in toluene (10 mL) at −35 °C. The reaction mixture was stirred for 1 h at 25 °C, producing a dark purple suspension. The solid was separated through filtration and the purple solution was collected. All volatiles were removed under reduced pressure producing a dark purple solid. The solid was rinsed with n-pentane and dried under reduced pressure to afford **2a** as a purple solid (561 mg, 865 µmol, 89%). **HR EI** (pos. mode): m/z calcd.

Results and Discussion

for ([$C_{30}H_{44}O_3{}^{35}Cl_1{}^{98}Mo_1{}^{32}S_2$]$^+$, [(OSSO)MoOCl]$^+$): 649.1469; found: 649.1486. **IR** in KBr: ($\tilde{v}$, cm^{-1}) 2956 (s), 2906 (m), 2869 (m), 1592 (vw), 1463 (m), 1433 (s), 1400 (m), 1361 (m), 1200 (m), 1252 (s), 1240 (s), 1202 (m), 1173 (m), 1139 (w), 1102 (m), 1022 (vw), 965 (vs, M=O), 916 (w), 880 (m), 837 (s), 774 (w), 769 (m), 748 (w), 658 (w), 649 (w), 559 (m), 499 (w), 486 (w), 417 (vw). **Elemental Analysis**: Calcd. for $C_{30}H_{44}ClO_3S_2Mo$: C 55.59, H 6.84; found: C 51.51, H 7.27. **EPR** (10 mM in THF, frequency: 9.4237 GHz, modulation amplitude: 0.150 mT, attenuation: 30 dB, sweep time: 60 s; B_0 sweep: 50 mT, B_0: 347 mT): g_{iso} = 1.94, A_{iso} = 130 MHz. **CV** (in DCM, **2a** (2 mM), [^{n}Bu$_4$N][PF$_6$] (100 mM), 100 mV s^{-1}): $E_{1/2}$ (1) = 0.57 V (reversible), $E_{1/2}$ (2) = −1.21 V (quasi-reversible). **Magnetic Properties**: μ_{eff} = 1.8 μ_B (Evans method, 298 K, DCM).

Chloro(oxo){1,2-dithiabutanediyl-2,2'-bis(4-methyl-6-tert-butyl-phenolato)}-molybdenum(V) (2b)

A solution of MoOCl$_3$ (44.1 mg, 202 μmol, 1.00 equiv.) in THF (5 mL) was added dropwise to a suspension of [(OSSO)K$_2$] (100 mg, 202 μmol, 1.00 equiv.) in THF (10 mL) at −78 °C. The suspension was stirred for 18 h at 25 °C during which the color changed from brown to purple. All volatiles were removed under reduced pressure producing a purple solid. The solid was extracted with toluene. All volatiles were removed under reduced pressure. The solid was rinsed with n-pentane and extracted with DCM. Removal of all volatiles under reduced pressure afforded **2b** as a purple solid (82.7 mg, 147 μmol, 73%). Single crystals suitable for X-ray diffraction were grown from a solution of **2b** in toluene, layered with n-hexane. **HR EI** (pos. mode): m/z calcd. for ([$C_{24}H_{32}O_3{}^{35}Cl_1{}^{98}Mo_1{}^{32}S_2$]$^+$, [(OSSO)MoOCl]$^+$): 565.05300; found: 565.05342. **IR** in KBr: ($\tilde{v}$, cm^{-1}) 3372 (w), 2993 (w), 2957 (m), 2920 (m), 2867 (m), 1762 (vw), 1594 (vw), 1481 (w), 1459 (m), 1431 (vs), 1403 (m), 1385 (m), 1360 (m), 1326 (w), 1282 (m), 1251 (s), 1229 (vs), 1210 (m), 1166 (m), 1125 (m), 1100 (m), 1024 (vw), 967 (vs, Mo=O), 936 (m), 915 (w), 861 (m), 831 (s), 814 (vs), 781 (vw), 759 (m), 739 (vw), 649 (m), 628 (m), 593 (vw), 565 (m), 555 (m), 538 (w), 476 (vw), 458 (w), 449 (w). **Elemental Analysis**: Calcd. for $C_{24}H_{32}ClO_3S_2Mo$: C 51.11.97, H 5.72; found: C 53.59, H 5.63. **EPR** (10 mM in THF, frequency: 9.4290 GHz, modulation amplitude: 0.150 mT, attenuation: 30 dB, sweep time: 60 s; B_0 sweep: 50 mT, B_0: 347 mT): g_{iso} = 1.94, A_{iso} = 130 MHz. **CV** (in DCM, **2b** (2 mM), [^{n}Bu$_4$N][PF$_6$] (100 mM), 250 mV s^{-1}): $E_{1/2}$ (1) = 0.48 V (reversible), $E_{1/2}$ (2) = −1.25 V (quasi-reversible). **Magnetic Properties**: μ_{eff} = 1.5 μ_B (Evans method, 298 K, DCM).

Group 6 Metal Complexes Featuring a Tetradentate (OSSO)-Type Ligand

rac-Chloro(oxo){dithiocyclohexanediyl-2,2'-bis(4,6-di-tert-butyl-phenolato)} molybdenum(V) (2c)

The synthesis of **2c** was performed analogously to **2a** using method A. The respective deprotonated pro-ligand [(OSSO)K$_2$] (50.0 mg, 79.0 μmol, 1.00 equiv.) and MoOCl$_4$ (20.0 mg, 79.0 μmol, 1.00 equiv.) were used. The complex was purified through extraction with *n*-hexane. All volatiles were removed under reduced pressure to afford **2c** as a purple/brown solid. **IR** in KBr: ($\tilde{v}$, cm^{-1}) 3356 (m), 2955 (vs), 2866 (m), 1589 (vw), 1462 (m), 1435 (s), 1400 (m), 1362 (m), 1284 (m), 1242 (s), 1203 (m), 1180 (m), 1138 (w), 1103 (m), 1022 (vw), 984 (vw), 957 (m, Mo=O), 914 (w), 883 (w), 837 (m), 760 (m), 744 (m), 644 (w), 613 (vw), 555 (m), 498 (w), 486 (w). **EI** (pos. mode): *m/z* calcd. for (C$_{34}$H$_{50}$ClMoO$_3$S$_2$$^+$): 703.2; found: 702.7.

rac-Chloro(oxo){dithiocyclohexanediyl-2,2'-bis(4,6-di-tert-butyl-phenolato)}tungsten(V) (4)

The synthesis of **4** was performed analogously to **2a** using method A. The respective potassium salt of the pro-ligand [(OSSO)K$_2$] (75.0 mg, 118 μmol, 1.00 equiv.) and WOCl$_4$ (40.5 mg, 118 μmol, 1.00 equiv.) were used. **4** was obtained as a purple/brown solid (48.1 mg, 60.9 μmol, 52%). **IR** in KBr: ($\tilde{v}$, cm^{-1}) 3368 (w), 2958 (vs), 2866 (m), 1462 (m), 1435 (m), 1396 (w), 1362 (m), 1284 (m), 1238 (m), 1203 (w), 1184 (w), 1138 (vw), 1103 (m), 968 (w, W=O), 918 (w), 879 (w), 868 (w), 845 (w), 756 (w), 675 (vw), 644 (vw), 555 (w), 494 (vw). **EI** (pos. mode): *m/z* calcd. for (C$_{34}$H$_{50}$ClWO$_3$S$_2$$^+$): 789.2; found: 788.1.

B.1.4.4 Catalytic Oxygen Atom Transfer Reactions

General procedure for reactions in DMSO-d_6

PPh$_3$ (266 μmol) and the catalyst (5 mol%, 5.3 μmol) were suspended in DMSO-d_6 (0.6 mL) and heated for 20 h at 80 °C. The conversion was determined through integration of the product and starting material resonances in the ^{1}H and ^{31}P{^{1}H} NMR spectra.[46]

General procedure for reactions in THF/THF-d_8

PPh$_3$ and the catalyst were dissolved in the solvent (0.6 mL). DMSO was added *via* a μL syringe to solution. The solutions were treated for the specified time at the specified temperatures. The conversion of PPh$_3$ to OPPh$_3$ was determined through integration of the starting material and product resonances as reported in literature.[20, 46]

Pre-oxidation of the catalyst

DMSO (106 μmol) was added to a solution of **1a** (5.3 μmol) in THF/THF-d_8 and the solution was heated for 2 h at 60 °C. During this time, the color of the solution changed from orange/red to purple. PPh$_3$ (106 μmol) was added as a solid to the purple solution and heating was continued for 2 h at 60 °C. The conversion of PPh$_3$ to OPPh$_3$ was determined through integration of the starting material and product resonances as reported in literature.[20, 46]

B.1.5 References

1. Kaim, W.; Schwederski, B., Biologische Funktion der "frühen" Übergangsmetalle: Molybdän, Wolfram, Vanadium, Chrom. In *Bioanorganische Chemie*, 4 ed.; Vieweg+Teubner Verlag: Wiesbaden, 2005; pp 222-247.
2. Schulzke, C.; Ghosh, A. C., Molybdenum and Tungsten Oxidoreductase Models. In *Bioinspired Catalysis: Metal-Sulfur Complexes*, Weigand, W.; Schollhammer, P., Eds. Wiley-VCH Verlag GmbH & Co. KGaA: Weinheim, Germany, 2014; pp 349-382.
3. Majumdar, A.; Sarkar, S. *Coord. Chem. Rev.* **2011**, *255*, 1039-1054.
4. Young, C. G. *J. Inorg. Biochem.* **2016**, *162*, 238-252.
5. Hagen, W. R. *Coord. Chem. Rev.* **2011**, *255*, 1117-1128.
6. Holm, R. H. *Coord. Chem. Rev.* **1990**, *100*, 183-221.
7. Ghosh, A. C.; Samuel, P. P.; Schulzke, C. *Dalton Trans.* **2017**, *46*, 7523-7533.
8. Williams, B. R.; Fu, Y.; Yap, G. P. A.; Burgmayer, S. J. N. *J. Am. Chem. Soc.* **2012**, *134*, 19584-19587.
9. Young, C. G. *J. Inorg. Biochem.* **2016**, *162*, 238-252.
10. Enemark, J. H.; Cooney, J. J. A.; Wang, J. J.; Holm, R. H. *Chem. Rev.* **2004**, *104*, 1175-1200.
11. Kaul, B. B.; Enemark, J. H.; Merbs, S. L.; Spence, J. T. *J. Am. Chem. Soc.* **1985**, *107*, 2885-2891.
12. Barnard, K. R.; Bruck, M.; Huber, S.; Grittini, C.; Enemark, J. H.; Gable, R. W.; Wedd, A. G. *Inorg. Chem.* **1997**, *36*, 637-649.
13. Hinshaw, C. J.; Peng, G.; Singh, R.; Spence, J. T.; Enemark, J. H.; Bruck, M.; Kristofzski, J.; Merbs, S. L.; Ortega, R. B.; Wexler, P. A. *Inorg. Chem.* **1989**, *28*, 4483-4491.
14. Spence, J. T.; Hinshaw, C. J. *Inorg. Chim. Acta* **1986**, *125*, L17-L21.
15. Gullotti, M.; Pasini, A.; Zanderighi, G. M.; Ciani, G.; Sironi, A. *J. Chem. Soc., Dalton Trans.* **1981**, *0*, 902-908.
16. Beckerle, K.; Sauer, A.; Spaniol, T. P.; Okuda, J. *Polyhedron* **2016**, *116*, 105-110.
17. Sauer, A. Early Transition Metal Bis(phenolate) Complexes for Selective Catalysis. Doctoral Thesis, RWTH Aachen University, Aachen, 2013.
18. Kaul, B. B.; Sellmann, D. *Z. Naturforsch.* **1983**, *38b*, 562-567.
19. Dilworth, J. R.; Richards, R. L.; Chen, G. J.-J.; Mcdonald, J. W., The Synthesis of Molybdenum and Tungsten Dinitrogen Complexes. In *Inorganic Syntheses*, Anelici, R. J., Ed. John Wiley & Sons, Inc.: Hoboken, NJ, USA, 1990; Vol. 28, pp 33-43.
20. Schindler, T.; Sauer, A.; Spaniol, T. P.; Okuda, J. *Organometallics* **2018**, *37*, 4336-4340.
21. Evans, D. F. *J. Chem. Soc.* **1959**, *0*, 2003-2005.
22. Evans, D. F.; Fazakerley, G. V.; Phillips, R. F. *J. Chem. Soc. A* **1971**, *0*, 1931-1934.
23. Atkins, P.; de Paula, J., *Physical Chemistry*. 9th ed.; Oxford University Press: Oxford, 2010.
24. Atagi, L. M.; Mayer, J. M. *Angew. Chem. Int. Ed.* **1993**, *32*, 439-441.
25. Elgrishi, N.; Rountree, K. J.; McCarthey, B. D.; Rountree, E. S.; Eisenhart, T. T.; Dempsey, J. L. *J. Chem. Educ.* **2018**, *95*, 197-206.
26. Nakamura, I.; Miras, H. N.; Fujiwara, A.; Fujibayashi, M.; Song, Y.-F.; Cronin, L.; Tsunashima, R. *J. Am. Chem. Soc.* **2015**, *137*, 6524-6530.
27. Larson, M. L.; Moore, F. W.; Edwards, D. A., Molybdenum Oxide Trichloride. In *Inorganic Syntheses*, Parry, R. W., Ed. John Wiley & Sons, Inc.: Hoboken, NJ, USA, 1970; Vol. 12, pp 190-192.
28. Behrens, A.; Behrens, U.; Nordlander, E.; Bader, C.; Rehder, D. *Inorg. Chim. Acta* **2005**, *358*, 1970-1974.
29. Mayer, J. M. *Inorg. Chem.* **1988**, *27*, 3899-3903.
30. Hagen, W. R. *Dalton Trans.* **2006**, *0*, 4415-4434.
31. Connelly, N. G.; Geiger, W. E. *Chem. Rev.* **1996**, *96*, 877-910.
32. Wong, Y.-L.; Yan, Y.; S. H. Chan, E.; Yang, Q.; C. W. Mak, T.; K. P. Ng, D. *J. Chem. Soc., Dalton Trans.* **1998**, *0*, 3057-3064.

33. Tsurugi, H.; Tanahashi, H.; Nishiyama, H.; Fegler, W.; Saito, T.; Sauer, A.; Okuda, J.; Mashima, K. *J. Am. Chem. Soc.* **2013**, *135*, 5986-5989.
34. Tanahashi, H.; Ikeda, H.; Tsurugi, H.; Mashima, K. *Inorg. Chem.* **2016**, *55*, 1446-1452.
35. Ma, X.; Schulzke, C.; Schmidt, H. G.; Noltemeyer, M. *Dalton Trans.* **2007**, *0*, 1773-1780.
36. Shimazaki, Y.; Stack, T. D.; Storr, T. *Inorg. Chem.* **2009**, *48*, 8383-8392.
37. Reddy, P. R.; Holm, R. H.; Caradonna, J. P. *J. Am. Chem. Soc.* **1988**, *110*, 2139-2144.
38. Lorber, C.; Plutino, M. R.; Elding, L. I.; Nordlander, E. *J. Chem. Soc., Dalton Trans.* **1997**, *2*, 3997-4003.
39. Tucci, G. C.; Donahue, J. P.; Holm, R. H. *Inorg. Chem.* **1998**, *37*, 1602-1608.
40. Nielson, A. J.; Andersen, R. A., Tungsten and Molybdenum Tetrachloride Oxides. In *Inorganic Syntheses*, Kirschner, S., Ed. John Wiley & Sons, Inc.: 1985; Vol. 23, pp 195-199.
41. Sousa, S. C. A.; Fernandes, A. C. *Coord. Chem. Rev.* **2015**, *284*, 67-92.
42. Marcseková, K.; Loos, C.; Rominger, F.; Doye, S. *Synlett.* **2007**, 2564-2568.
43. Beckerle, K.; Capacchione, C.; Ebeling, H.; Manivannan, R.; Mülhaupt, R.; Proto, A.; Spaniol, T. P.; Okuda, J. *J. Organomet. Chem.* **2004**, *689*, 4636-4641.
44. Capacchione, C.; Proto, A.; Ebeling, H.; Mülhaupt, R.; Möller, K.; Spaniol, T. P.; Okuda, J. *J. Am. Chem. Soc.* **2003**, *125*, 4964-4965.
45. Kruse, T.; Weyhermüller, T.; Wieghardt, K. *Inorg. Chim. Acta* **2002**, *331*, 81-89.
46. Fulmer, G. R.; Miller, A. J. M.; Sherden, N. H.; Gottlieb, H. E.; Nudelman, A.; Stoltz, B. M.; Bercaw, J. E.; Goldberg, K. I. *Organometallics* **2010**, *29*, 2176-2179.

Group 6 Metal Complexes Featuring a Tetradentate (OSSO)-Type Ligand

B.2 Mononuclear Complexes Featuring a Tris(ONNO)-Type Ligand

B.2.1 Introduction

The field of cyclic polyether compounds was shaped by the pioneering work of Pedersen who was awarded the Nobel prize for chemistry (jointly with D. J. Cram and J.-M. Lehn) in 1987 for his discovery of crown ethers in 1967.[1,2] The first macrocyclic crown ether pro-ligand dibenzo-18-crown-6 was synthesized from catechol and bis(2-chloroethyl) ether with sodium hydroxide as a base.[1] Since then, a large library of macrocyclic polyethers has been developed and a large number of alkaline and alkaline earth metal complexes stabilized by crown ether ligands have been reported.[3-5] When 18-crown-6 (18c6) is used to stabilize calcium complexes with weakly coordinating anions, it produces a calcium center with hexagonal bipyramidal coordination geometry featuring two solvent molecules in the apical positions (Figure B.2.1).[6]

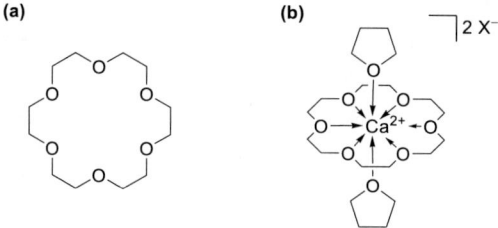

Figure B.2.1. (a) Macrocyclic polyether 18c6 pro-ligand and (b) dicationic calcium 18c6 complex featuring two non-coordinating anions (X⁻ = [B(Ph)$_3$(η^1-C$_3$H$_5$)]⁻).[1,6]

Compartmental ligands prepared by Reinhoudt and co-workers represent a further development of crown ether ligands.[7-9] These ligands combine a cyclic polyether compartment with a salen (ONNO)-type binding site to selectively stabilize heterodinuclear complexes. Several types of (ONNO)-crown-metallomacrocycles have been synthesized featuring various polyether chain lengths and diamine backbones. To synthesize these ligands, a barium cation acts as a template which is coordinated by a linear polyether chain featuring dialdehyde terminal groups. The templating cation brings the two chain ends conformationally close to each other, facilitating the ring-closing condensation reaction between the two aldehyde groups and a diamine to produce the corresponding Schiff base (Scheme B.2.1). The resulting mononuclear complex contains a vacant salen-type binding site which can be metalated with smaller transition metals to afford heterodinuclear complexes. In these complexes, the hard template cation interacts with all oxygen atoms of the crown ether-type cavity and the soft transition metal is exclusively situated in the salen-type binding site. Removal of the barium ion renders a mononuclear complex featuring a vacant crown ether-type compartment and an occupied salen-type binding site.

47

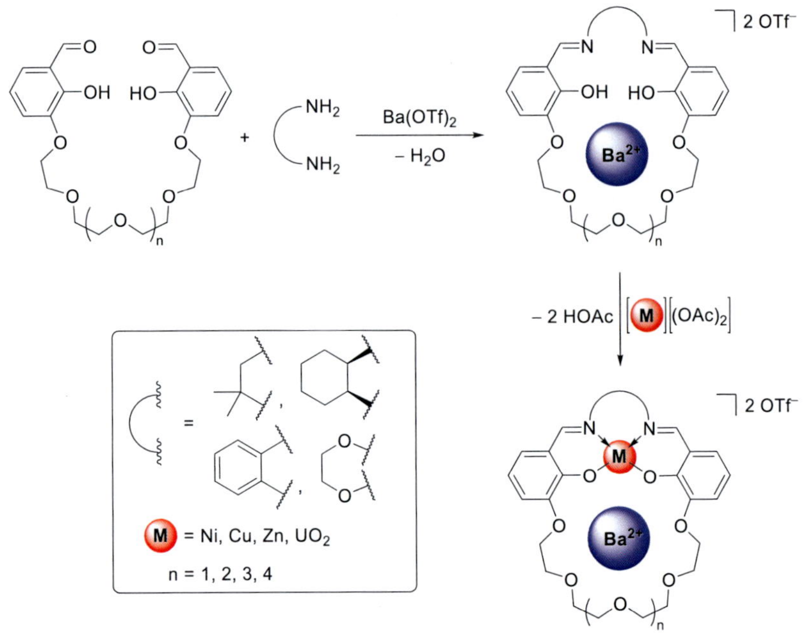

Scheme B.2.1. Template directed synthesis to produce heterodinuclear complexes featuring (ONNO)-crown-metallomacrocycles.[7-11]

Further research on compartmental ligands led to the development of tetranucleating macrocycles featuring a 18c6-type cavity and three vacant salen-type binding sites.[12] The tris(ONNO)-type ligand was synthesized by Reinhoudt and co-workers through pre-coordination of three equivalents of 2,3-dihydroxy-1,4-napthalenedialdehyde with a barium cation template. Subsequent addition of equimolar amounts of diamine produces the corresponding Schiff base (Scheme B.2.2). The phenolic OH groups of this ligand framework are intramolecularly bridged by hydrogen bonds to the respective imine bonds, producing a cavity that resembles 18c6 (Figure B.2.1). Selective formation of the [3+3] cyclization product is achieved with Ba^{2+} as templating cation and with C_2-bridging units such as ethylenediamine, trans-1,2-cyclohexanediamine or benzene-1,2-diamine (Scheme B.2.2). Performing the synthesis with more flexible bridging units such as 1,3-diaminopropane produces varying amounts of [3+3] and [4+4] macrocycles. Mixing equimolar amounts of dialdehyde and diamine without a template cation affords insoluble polymeric residues. However, the structural aspects of the mononuclear macrocycles have not been investigated and the precise molecular structure remains elusive since single crystal X-ray diffraction analysis has not yet been performed.

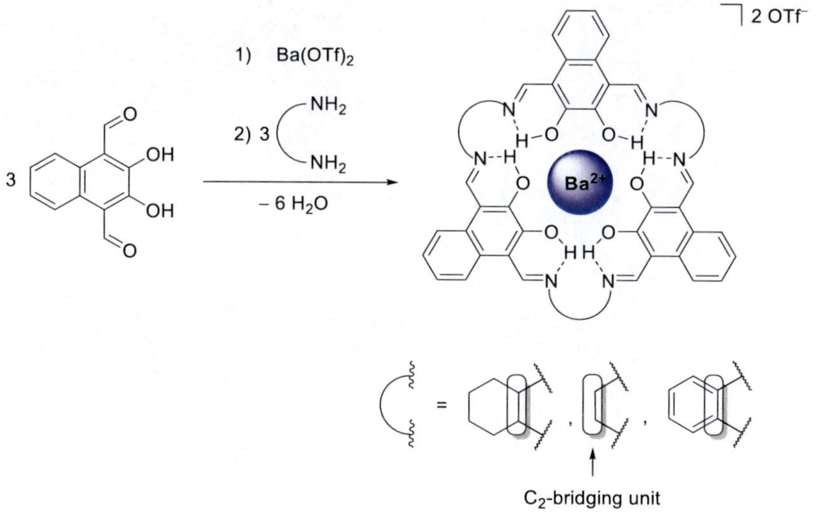

Scheme B.2.2. Synthesis of mononuclear tris(ONNO)-type macrocyclic complexes featuring C_2-bridging units.

MacLachlan and co-workers compared the structural properties of macrocycles based on 2,3-dihydroxy-1,4-napthalenedialdehyde to the ones based on 3,6-diformylcatechol (Chapter A.2).[13] Condensation of the dialdehyde with phenylenediamine derivatives selectively produces the [3+3] macrocycle featuring vacant salen-type and 18c6-type compartments. However, the spectroscopic properties of these macrocycles strongly deviate from the ones observed for macrocycles based on 3,6-diformylcatechol. Structural analysis of model compounds based on singular naphthalene subunits enabled observation of a keto enol tautomerization, rationalizing the observed spectroscopic properties. Theoretical calculations indicate that the macrocycle predominantly adopts the keto-enamine tautomer but exists as a mixture of the two tautomers. Tautomerization eliminates the conjugation in the macrocycle, decreasing its propensity of π-π-stacking and inhibiting the formation of single crystals suitable for X-ray diffraction analysis. The precise molecular structure of the pro-ligands remains elusive similar to the mononuclear Ba^{2+} complexes.[12, 13]

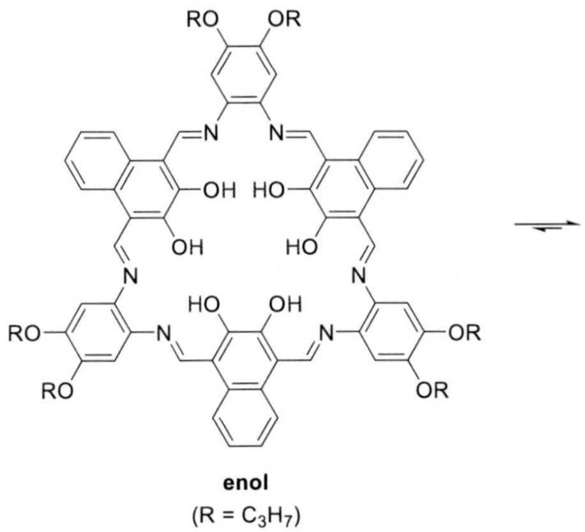

Mononuclear Complexes Featuring a Tris(ONNO)-Type Ligand

enol
(R = C$_3$H$_7$)

keto
(R = C$_3$H$_7$)

Scheme B.2.3. Tautomerization of the [3+3] macrocycle based on 2,3-dihydroxy-1,4-napthalenedialdehyde.[13]

B.2.2 Results and Discussion

Heterometallic core/shell template synthesis has been commonly employed to selectively produce deprotonated [3+3] macrocycles based on 3,6-diformylcatechol, avoiding undesired oligomerization reactions.[12, 14, 15] However, removal of the template cations to produce the corresponding protonated pro-ligands is challenging due to reversible Schiff base rearrangements.[16] Template removal is only successful with macrocycles featuring inert oxime bonds which inhibit C=N bond recombinations.[16, 17] Inspired by the results by Reinhoudt and co-workers on protonated ligands featuring vacant salen-type but occupied 18c6-type binding sites, mononuclear complexes featuring tris(ONNO)-type ligands based on 3,6-diformylcatechol were synthesized.[12] Alkaline earth metal and lanthanide metal complexes were prepared, and their structural properties assessed. Protonated mononuclear complexes featuring tris(ONNO)-type ligands may be sufficiently stabilized by the template cation against Schiff base rearrangements to give access to heterometallic complexes in a subsequent step.

The mononuclear complex $[H_6^{Cy}LCa(OTf)_2]$ (**[6$_{Ca}$][(OTf)$_2$]**) was prepared in a similar approach to the one reported by Reinhoudt and co-workers.[12] **[6$_{Ca}$][(OTf)$_2$]** was prepared through pre-coordination of the Ca^{2+} template to 3,6-diformylcatechol in a boiling equivolumic mixture of acetonitrile/methanol (MeCN/MeOH, 1:1) to produce a pale yellow solution. Subsequent condensation of the dialdehyde with an excess of *rac-trans*-1,2-diaminocyclohexane facilitated the cyclization reaction to produce **[6$_{Ca}$][(OTf)$_2$]** (Scheme B.2.4). The monometallic complex was quantitatively obtained as a red solid and purified through rinsing with Et_2O and *n*-hexane to remove excess diamine.

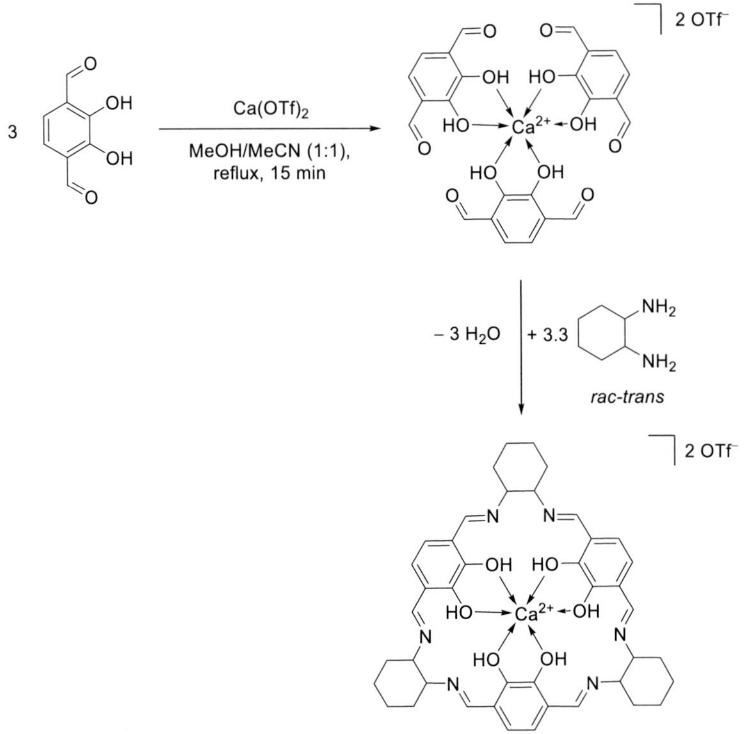

Mononuclear Complexes Featuring a Tris(ONNO)-Type Ligand

Scheme B.2.4. Two-step template synthesis of the mononuclear complex [6$_{Ca}$][(OTf)$_2$].

Removal of all volatile solvents is challenging due to coordination of solvent molecules to the metal center and enclosure of solvent molecules in the lattice of the crystallites. [6$_{Ca}$][(OTf)$_2$] is soluble in DMSO, DMF and MeOH/MeCN (1:1) but only sparingly soluble in MeOH or MeCN alone. Drying under reduced pressure partially removes ancillary MeOH ligands, rendering it insoluble in pure MeCN. Upon addition of methanol, the complex becomes again soluble in MeCN. The complex is insoluble in THF, DCM and toluene. The FAB mass spectrum of [6$_{Ca}$][(OTf)$_2$] shows a prominent peak at m/z = 921.3 which can be assigned to ([[6$_{Ca}$][(OTf)$_2$] − OTf⁻]⁺). The IR spectrum indicates complete consumption of the dialdehyde carbonyl at $\tilde{v}$ = 1663 cm⁻¹ in favor of the imine bond with an absorption at $\tilde{v}$ = 1632 cm⁻¹ (Figure B.2.2).[15] The IR spectrum also shows the characteristic absorptions of the triflate anion at $\tilde{v}$ = 1350 − 1200 cm⁻¹ ($v_{as}(SO_3)$), 1031 cm⁻¹ ($v_s(SO_3)$) and 638 cm⁻¹ ($\delta_s(SO_3)$).[18]

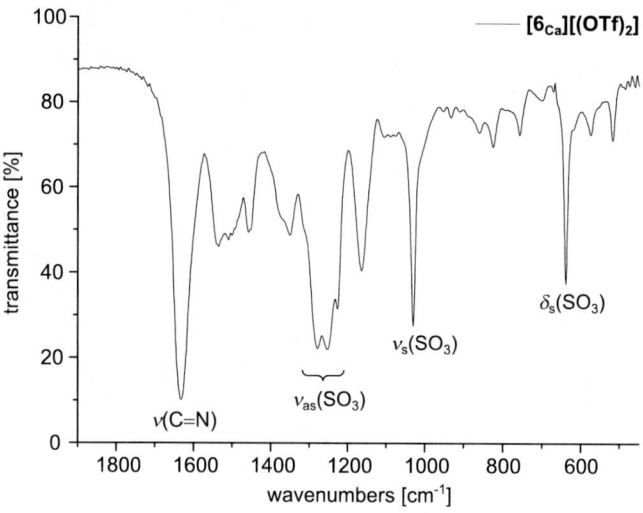

Figure B.2.2. Excerpt of the IR spectrum of [6$_{Ca}$][(OTf)$_2$] measured on a KBr pellet.

The ^{1}H NMR spectrum of [6$_{Ca}$][(OTf)$_2$] in DMSO-d_6 shows three resonances for the imine and aromatic protons at δ 8.5 – 8.3 ppm and at δ 6.8 – 6.7 ppm (Figure B.2.3). The respective imine and aromatic resonances did not coalesce up to a temperature of 80 °C, indicating that they are produced by diastereotopic protons rather than by an exchange process. The three resonances are attributed to the formation of three diastereomers (R,R-R,R-R,R/S,S-S,S-S,S), (R,R-S,S-R,R) and (S,S-R,R-S,S). The ^{1}H NMR spectrum of [6$_{Ca}$][(OTf)$_2$] in DMSO-d_6 also revealed two resonances for the phenolic protons at δ 13.3 – 12.8 ppm, indicating retention of the protonated state. Since separation of the diastereomers and enantiomers is challenging, the enantiopure diamines (R,R)-trans-1,2-diaminocyclohexane and (S,S)-trans-1,2-diaminocyclohexane were used in stochiometric amounts to prepare the complexes with all (R,R-R,R-R,R) and (S,S-S,S-S,S) configuration. The enantiopure complexes (R,R)-[H$_6^{Cy}$LCa(OTf)$_2$] ([7$_{Ca}$][(OTf)$_2$]) and (S,S)-[H$_6^{Cy}$LCa(OTf)$_2$] ([8$_{Ca}$][(OTf)$_2$]), featuring all (R,R-R,R-R,R) and all (S,S-S,S-S,S) configuration, were prepared similar to [6$_{Ca}$][(OTf)$_2$] and obtained in high yield. The ^{1}H NMR spectra of [7$_{Ca}$][(OTf)$_2$] and [8$_{Ca}$][(OTf)$_2$] in DMSO-d_6 show one resonance for the imine protons at δ 8.44 ppm and one for the aromatic protons at δ 6.74 ppm (Figure B.2.3). In the ^{13}C{^{1}H} NMR spectrum of [7$_{Ca}$][(OTf)$_2$] and [8$_{Ca}$][(OTf)$_2$] in DMSO-d_6 only one resonance for the imine carbon (HC=N) is observed at δ 161.8 ppm, whereas in case of [6$_{Ca}$][(OTf)$_2$] two sets of signals were observed at δ 161.8 and 165.4 ppm, respectively. The phenolic protons of [7$_{Ca}$][(OTf)$_2$] and [8$_{Ca}$][(OTf)$_2$] give a broad resonance at δ 13.6 – 11.5 ppm. The NMR spectroscopic data confirms the formation of a D_3 symmetric complex in which all [3+3] subunits are equivalent. The resonances were assigned by NOESY, HSQC and COSY spectra. The triflate anions exhibit a singlet resonance at δ −77 ppm in the ^{19}F{^{1}H} NMR spectrum that indicates non-coordinating behavior in solution. In case of [7$_{Ca}$][(OTf)$_2$] and [8$_{Ca}$][(OTf)$_2$], the absorption of the imine bond is shifted to $\tilde{\nu}$(C=N) = 1637 cm^{-1}.

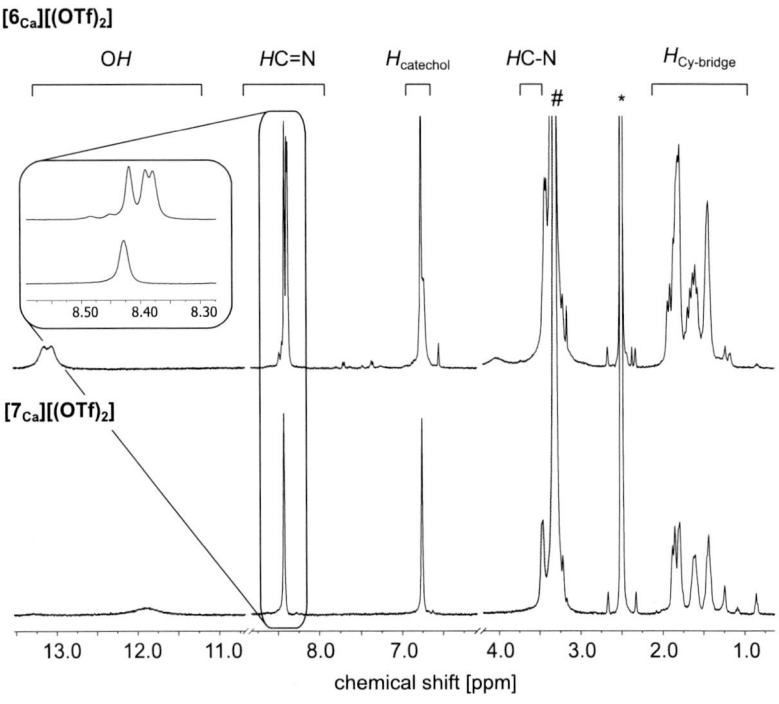

Figure B.2.3. Comparison of the ^{1}H NMR spectra of [6$_{Ca}$][(OTf)$_2$] and [7$_{Ca}$][(OTf)$_2$] in DMSO-d_6, referenced to residual (*) DMSO-d_5 with (#) H$_2$O.

To elucidate which metal cations facilitate selective formation of the [3+3] macrocycle, other di- and tricationic metals with similar ionic radii and coordination environments to hexacoordinate Ca^{2+} (ionic radius: 1.00 Å) were used.[19] Since Reinhoudt and co-workers used Ba^{2+} as a template cation, which has an ionic radius of 1.35 Å, metal cations with ionic radii between the ones of Ca^{2+} and Ba^{2+} (1.00 – 1.35 Å) were selected.[12, 19] The alkaline earth metals Sr^{2+} and Ba^{2+} and the lanthanides La^{3+} and Ce^{3+} selectively facilitate the [3+3] cyclization reaction. Sr(OTf)$_2$, Ba(OTf)$_2$, La(OTf)$_3$ and CeCl$_3$ were used to prepare the corresponding mononuclear complexes [7$_{Sr}$][(OTf)$_2$], [7$_{Ba}$][(OTf)$_2$], [7$_{La}$][(OTf)$_3$] and [7$_{Ce}$][Cl$_3$] similar to [7$_{Ca}$][(OTf)$_2$] (Scheme B.2.5).

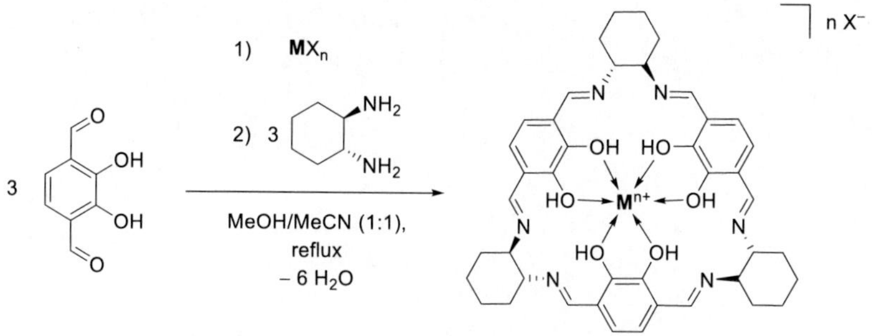

Results and Discussion

Scheme B.2.5. Synthesis of the monocular complexes [7$_{Sr}$][(OTf)$_2$], [7$_{Ba}$][(OTf)$_2$], [7$_{La}$][(OTf)$_3$] and [7$_{Ce}$][Cl$_3$].

MX$_n$	Complex
MX$_n$ = Sr(OTf)$_2$:	[7$_{Sr}$][(OTf)$_2$]
MX$_n$ = Ba(OTf)$_2$:	[7$_{Ba}$][(OTf)$_2$]
MX$_n$ = La(OTf)$_3$:	[7$_{La}$][(OTf)$_3$]
MX$_n$ = CeCl$_3$:	[7$_{Ce}$][Cl$_3$]

The IR spectra of the complexes show absorptions in the range of $\tilde{v}$ = 1630 – 1637 cm^{-1}, which correspond to the imine stretching modes and indicate completion of the Schiff base reaction. The mononuclear complexes were quantitatively obtained as red/orange solids in case of [7$_{Sr}$][(OTf)$_2$] and [7$_{Ba}$][(OTf)$_2$] and as dark wine-red solids in case of [7$_{La}$][(OTf)$_3$] and [7$_{Ce}$][Cl$_3$]. The FAB mass spectra of these complexes show peaks at m/z = 968.3 ([7$_{Sr}$][(OTf)$_2$]), 1018.1 ([7$_{La}$][(OTf)$_3$]) and 907.2 ([7$_{Ce}$][Cl$_3$]), which can be assigned to the [M – X]$^-$ fragments in case of [7$_{Sr}$][(OTf)$_2$] and to the [M – X$^-$, – HX]$^+$ fragments in case of [7$_{La}$][(OTf)$_3$] and [7$_{Ce}$][Cl$_3$]. The FAB mass spectrum of [7$_{Sr}$][(OTf)$_2$] is given as an example in Figure B.2.4.

Mononuclear Complexes Featuring a Tris(ONNO)-Type Ligand

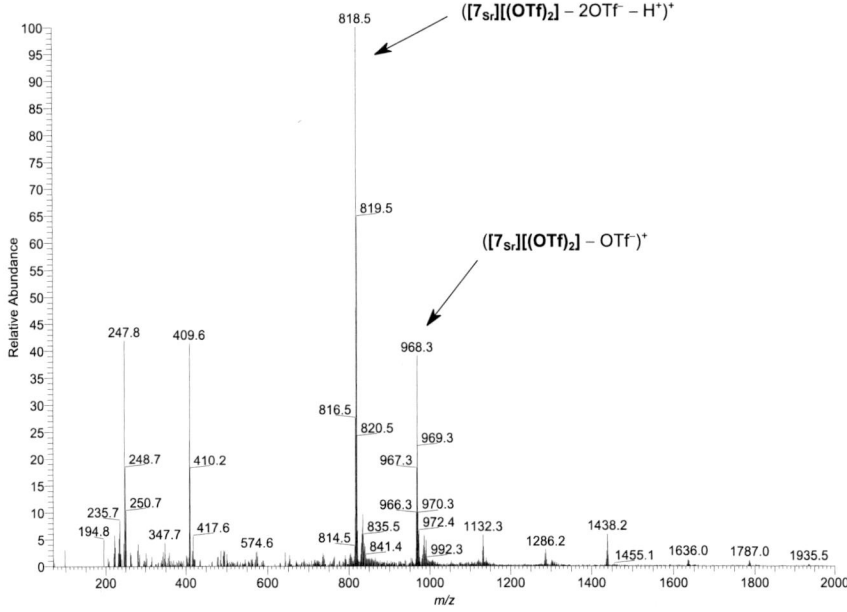

Figure B.2.4. FAB mass spectrum of **[7$_{Sr}$][(OTf)$_2$]** in 3-NBA matrix, measured in positive mode.

Due to the low solubility of the complexes in DMSO, the NMR spectroscopic characterization was performed using an equivolumic mixture of MeCN-d_3 and MeOD-d_4. The ^{1}H NMR spectra of **[7$_{Ca}$][(OTf)$_2$]**, **[7$_{Sr}$][(OTf)$_2$]**, **[7$_{Ba}$][(OTf)$_2$]** and **[7$_{La}$][(OTf)$_3$]** in MeCN-d_3/MeOD-d_4 (1:1) feature one resonance for the imine protons at δ 8.47 – 8.53 ppm and one for the aromatic protons at δ 6.64 – 6.76 ppm, indicating small influence of the central cation onto the chemical shifts of the resonances (Figure B.2.5). In case of **[7$_{Ce}$][Cl$_3$]**, the resonances of the imine protons and the ones of the aromatic protons are downfield shifted to δ 10.51 ppm and δ 8.31 ppm, respectively. Similarly, the signal of the imine carbon (HC=N) is shifted by 3.6 ppm to δ 167.7 ppm in the ^{13}C{^{1}H} NMR spectrum. Since cerium is in the formal oxidation state +III, the unpaired electron and, thus, paramagnetic nature of the complex account for the large shifts of the resonances in the ^{1}H and ^{13}C{^{1}H} NMR spectra of **[7$_{Ce}$][Cl$_3$]**.

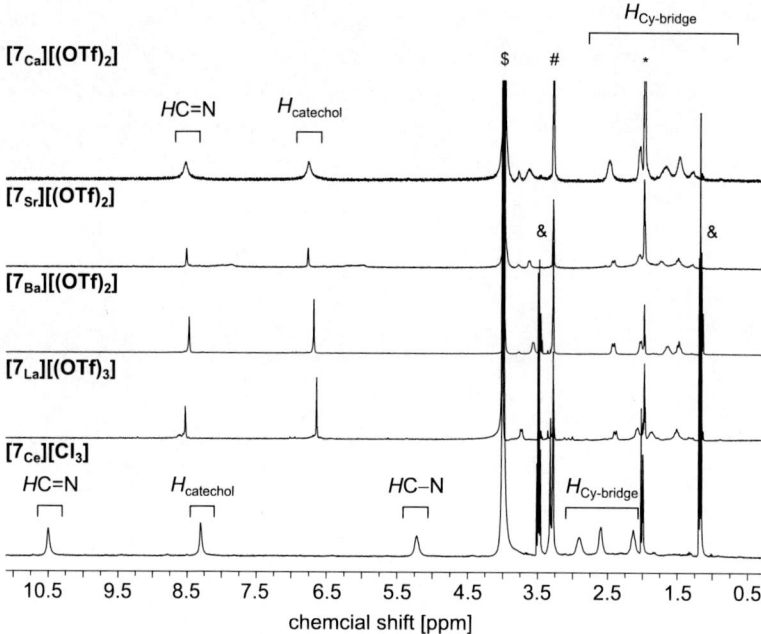

Figure B.2.5. Comparison of the ^{1}H NMR spectra of [7$_{Ca}$][(OTf)$_2$], [7$_{Sr}$][(OTf)$_2$], [7$_{Ba}$][(OTf)$_2$], [7$_{La}$][(OTf)$_3$] and [7$_{Ce}$][Cl$_3$] in MeCN-d_3/MeOD-d_4 referenced to residual solvent signals (#) MeOD-d_3 and (*) MeCN-d_2 with traces of ($) H$_2$O and (&) Et$_2$O.

Due to the limited solubility of the complexes featuring the 1,2-cyclohexanediamine bridging unit, other substituted diamines such as (R,R)-1,2-diphenylethylenediamine and 2,2-dimethylpropane-1,3-diamine were tested in the cyclization reaction. Using 2,2-dimethylpropane-1,3-diamine and Ca(OTf)$_2$ as template, selective formation of the [3+3] macrocycle was not observed. Instead a mixture of two macrocycles with different ring sizes is produced, indicated by two resonances of the methyl protons in the ^{1}H NMR spectrum of the reaction mixture in DMSO-d_6 at δ 1.60 and 1.11 ppm, respectively. The FAB mass spectrum confirms the formation of two macrocycles with m/z = 652.6 ([2+2] macrocycle) and 1116.5 ([4+4] macrocycle), which correspond to the [M − OTf$^-$]$^+$ fragments. Reinhoudt and co-workers also reported the formation of a [4+4] macrocycle when performing the reaction with 2,3-dihydroxy-1,4-napthalenedialdehyde and the C$_3$-bridging unit 1,3-propanediamine, affording a mixture of the respective [3+3] and [4+4] macrocycles.[12]

Mononuclear Complexes Featuring a Tris(ONNO)-Type Ligand

Using the C_2-bridging unit (R,R)-1,2-diphenylethylenediamine, [3+3] macrocyclic complexes were selectively obtained. **[9$_{Ca}$][(OTf)$_2$]**, **[9$_{Sr}$][(OTf)$_2$]**, **[9$_{Ba}$][(OTf)$_2$]**, **[9$_{La}$][(OTf)$_3$]** and **[9$_{Ce}$][Cl$_3$]** were prepared similar to **[6$_{Ca}$][(OTf)$_2$]** through pre-coordination of the metal cations by three equivalents of 3,6-diformylcatechol in boiling MeCN/MeOH (1:1) and subsequent condensation of the dialdehyde with equimolar amounts of (R,R)-1,2-diphenylethylenediamine (Scheme B.2.6).

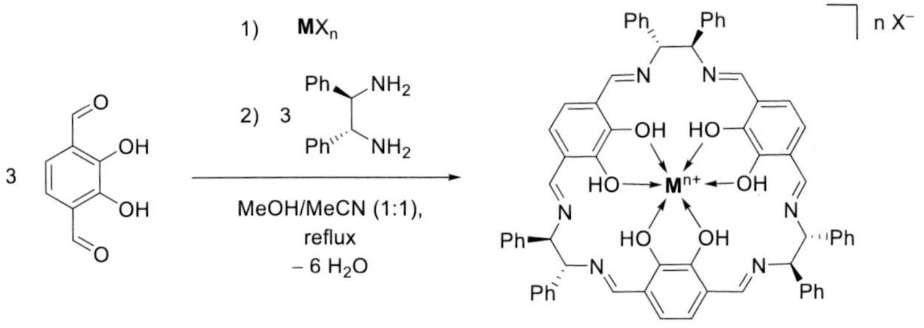

MX$_n$ = Ca(OTf)$_2$: **[9$_{Ca}$][(OTf)$_2$]**

MX$_n$ = Sr(OTf)$_2$: **[9$_{Sr}$][(OTf)$_2$]**

MX$_n$ = Ba(OTf)$_2$: **[9$_{Ba}$][(OTf)$_2$]**

MX$_n$ = La(OTf)$_3$: **[9$_{La}$][(OTf)$_3$]**

MX$_n$ = CeCl$_3$: **[9$_{Ce}$][Cl$_3$]**

Scheme B.2.6. Synthesis of the monocular complexes **[9$_{Ca}$][(OTf)$_2$]**, **[9$_{Sr}$][(OTf)$_2$]**, **[9$_{Ba}$][(OTf)$_2$]**, **[9$_{La}$][(OTf)$_3$]** and **[9$_{Ce}$][Cl$_3$]**.

The IR spectra of the mononuclear complexes confirm completion of the Schiff base reactions as indicated by absorptions in the range of $\tilde{v}$ = 1632 – 1621 cm^{-1}. Similar to the analogous complexes featuring the 1,2-cyclohexanediamine bridging unit, the complexes were quantitatively obtained as red/orange solids in case of **[9$_{Ca}$][(OTf)$_2$]**, **[9$_{Sr}$][(OTf)$_2$]**, **[9$_{Ba}$][(OTf)$_2$]** and as dark wine-red solids in case of **[9$_{La}$][(OTf)$_3$]** and **[9$_{Ce}$][Cl$_3$]**. The FAB mass spectra of the complexes show prominent peaks at m/z = 1215.6 (**[9$_{Ca}$][(OTf)$_2$]**), 1262.1 (**[9$_{Sr}$][(OTf)$_2$]**) and 1313.3 (**[9$_{Ba}$][(OTf)$_2$]**), which correspond to the [M – OTf$^-$]$^+$ fragments, and at m/z = 1312.0 (**[9$_{La}$][(OTf)$_3$]**) and 1200.3 (**[9$_{Ce}$][Cl$_3$]**), which correspond to the [M – X$^-$, – HX]$^+$ fragments. The FAB mass spectrum of **[9$_{Sr}$][(OTf)$_2$]** is given as an example in Figure B.2.6.

Results and Discussion

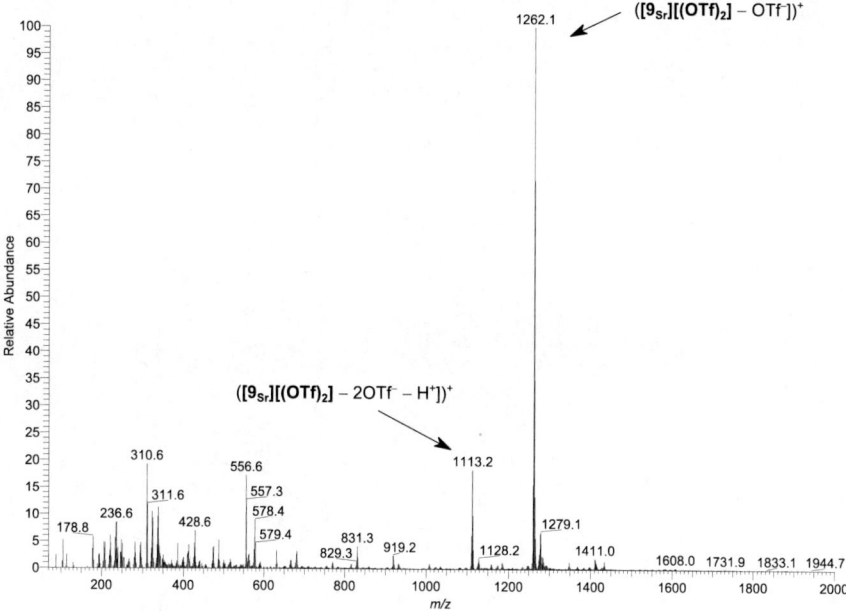

Figure B.2.6. FAB mass spectrum of [9$_{Sr}$][(OTf)$_2$] in 3-NBA matrix, measured in positive mode.

The ^{1}H NMR spectra of the complexes in MeCN-d_3/MeOD-d_4 (1:1) feature only one set of signals for the imine protons and the aromatic protons of the catechol subunit, indicating selective formation of the D_3 symmetric complexes. Similar to the complexes with the 1,2-cyclohexanediamine bridging unit, the template cation only exhibits a small influence onto the chemical shifts of the imine protons at δ 8.18 – 8.25 ppm and the aromatic protons of the catechol subunit at δ 6.41 – 6.55 ppm. The resonances of the imine protons and aromatic protons of the catechol subunit of [9$_{Ce}$][Cl$_3$] are again downfield shifted to δ 9.98 and 7.94 ppm, respectively.

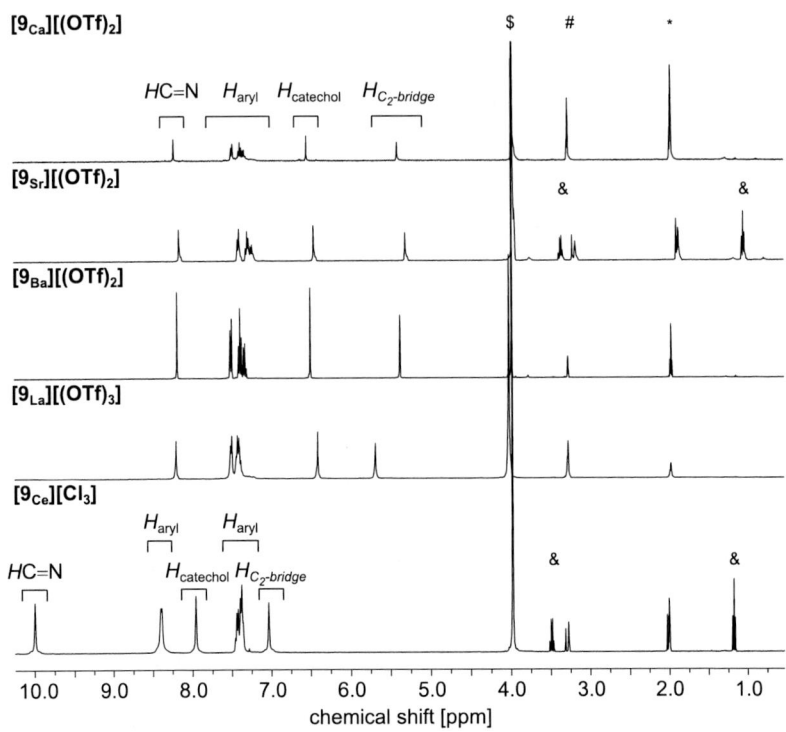

Figure B.2.7. Comparison of the ¹H NMR spectra of [9_Ca][(OTf)$_2$], [9_Sr][(OTf)$_2$], [9_Ba][(OTf)$_2$], [9_La][(OTf)$_3$] and [9_Ce][Cl$_3$] in MeCN-d_3/MeOD-d_4 referenced to residual solvent signals (#) MeOD-d_3 and (*) MeCN-d_2 with traces of ($) H$_2$O and (&) Et$_2$O.

Single crystals of [9_Ca][(OTf)$_2$] were grown from MeCN/MeOH (1:1) layered with Et$_2$O and *n*-hexane and the molecular structure was determined by X-ray diffraction analysis (Figure B.2.8(a)). Single crystals of [9_Sr][(OTf)$_2$] and [9_Ba][(OTf)$_2$] were also grown from MeCN/MeOH (1:1) through vapor diffusion of Et$_2$O into the solutions. Due to phase transition at low temperature, the diffraction data could not be solved, and the molecular structures could not be obtained. Due to disordered solvent molecules, the diffraction data of [9_Ca(MeOH)$_2$][(OTf)$_2$] was treated with the SQUEEZE routine.[20] The central calcium cation is coordinated in the 18c6-type cavity in distorted hexagonal bipyramidal fashion featuring the six catechol oxygen atoms in the equatorial plane and two molecules of methanol in the apical positions (Figure B.2.8(b), (c)). The coordination geometry of the central calcium atom is similar to the ones reported for other mononuclear calcium complexes coordinated by 18c6 featuring non-coordinating anions.[6,21] The methanol ligands further substantiate the previously observed solvation effects. As expected, the molecular structure contains vacant salen-type binding sites and non-coordinating triflate anions. The three catechol subunits are tilted in such a way that the complex adopts a drilled propeller-like structure with Δ configuration. The C–O (1.286(6) – 1.351(6) Å) and C=N bond distances (1.271(6) – 1.294(6) Å) are similar to other enol-imine tautomers of tris(ONNO)-type ligands (C–O (1.328 1.370) and C=N (1.224 – 1.298 Å)), rendering keto-enol tautomerization unlikely.[13, 22]

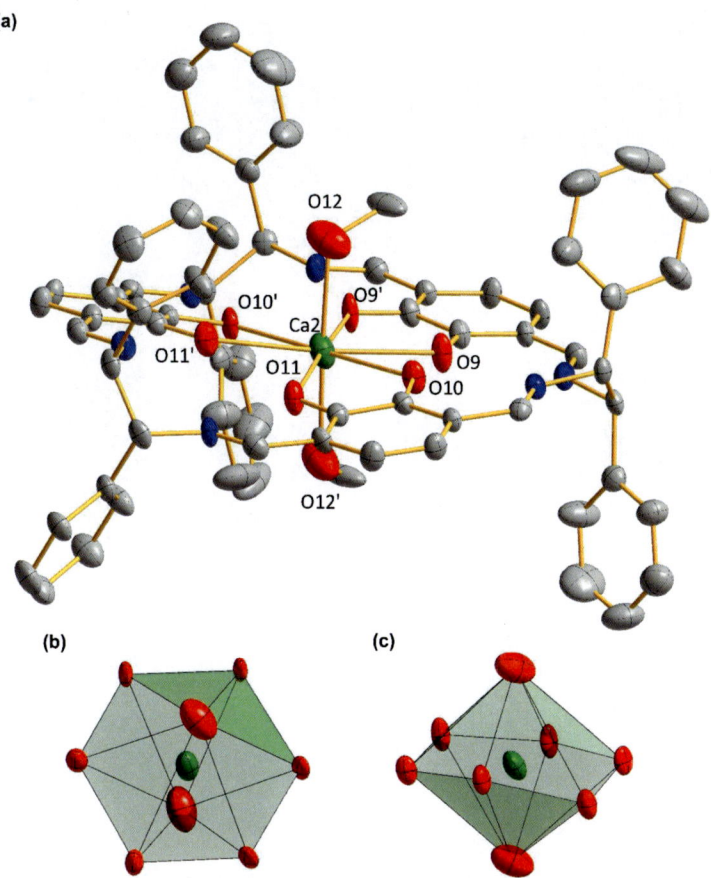

Figure B.2.8. (a) Molecular structure of **[9$_{Ca}$(MeOH)$_2$][(OTf)$_2$]** with 50% displacement ellipsoids, all hydrogen atoms and the triflate anions are omitted for clarity. Coordination geometry of the calcium atom in complex **[9$_{Ca}$(MeOH)$_2$][(OTf)$_2$]**, (b) top view of the polyhedron and (c) side view of the polyhedron, only atoms coordinated to calcium are shown (oxygen: red, calcium: green).

A variable temperature ¹H NMR experiment of **[9_Ca][(OTf)_2]** in MeCN-d_3/MeOD-d_4 was performed to investigate whether the complex rapidly exchanges between the Δ- and Λ isomer in solution. The signals of the imine protons and aromatic protons do not split into two sets of signals at lower temperatures. The NMR experiment indicates that the complex either exclusively adopts the Δ form in solution or that interconversion between the Δ and Λ isomer remains rapid. In the ¹H NMR spectra of **[9_Ca][(OTf)_2]** in DMSO-d_6, a minor second species is observed. The second species produces resonances at δ 6.64 and 5.56 ppm that can be assigned to the protons of the catechol subunit and the ethylene protons of the C_2-bridging unit. A ROESY experiment revealed a positive exchange process between the major and minor species which may be attributed to interconversion between the Δ and Λ isomers of the complex (Figure B.2.9). Presumably, the exchange process is less rapid in DMSO compared to MeCN/MeOH which enabled detection on the NMR time scale.

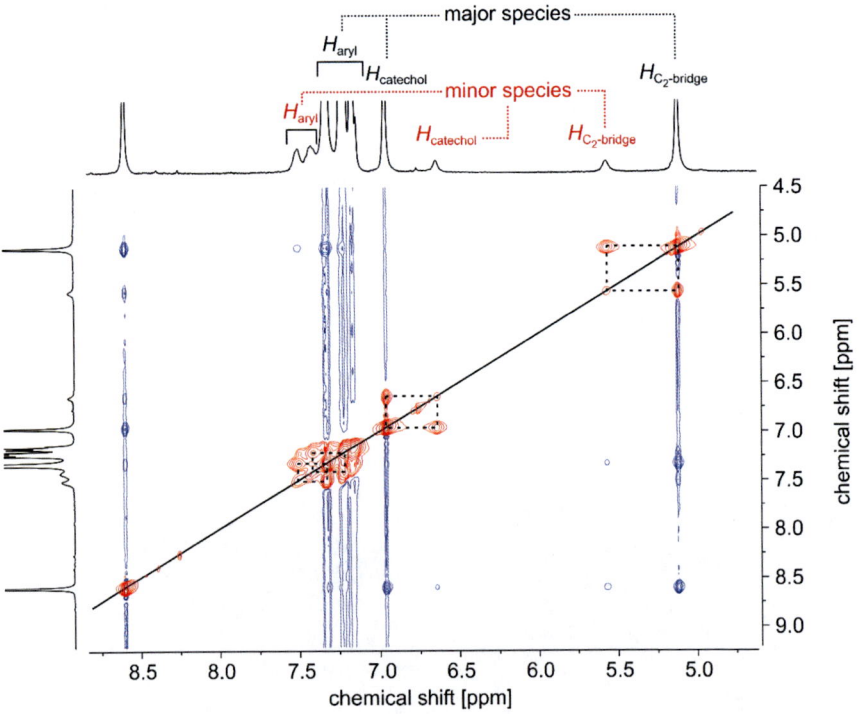

Figure B.2.9. ROESY experiment of **[9Ca][(OTf)_2]** in DMSO-d_6 with indicated exchange interaction between the major and minor species.

Due to the conjugated macrocyclic ligand framework, all prepared mononuclear complexes are brightly colored. The UV-Vis spectrum of **[9$_{La}$][(OTf)$_3$]** in MeCN/MeOH (1:1) exhibits an absorption maximum at 347 nm (C=N, $\pi \rightarrow \pi^*$) which is shifted to higher wavelengths compared to the alkaline earth metal complex **[9$_{Sr}$][(OTf)$_2$]** with a maximum at 310 nm (Figure B.2.10). A weaker absorption band is observed at 506 nm (C=N, $n \rightarrow \pi^*$) for **[9$_{La}$][(OTf)$_3$]** and at 471 nm for **[9$_{Sr}$][(OTf)$_2$]**. Both absorption bands of **[9$_{La}$][(OTf)$_3$]** are shifted by approximately 36 nm to higher wavelengths. The red-shift of the absorptions account for the wine-red color of **[9$_{La}$][(OTf)$_3$]** compared to the orange/red color of **[9$_{Sr}$][(OTf)$_2$]**. The absorption maxima of **[9$_{Sr}$][(OTf)$_2$]** and **[9$_{La}$][(OTf)$_3$]** are red-shifted to higher wavenumbers compared to non-coordinating pro-ligands.[23]

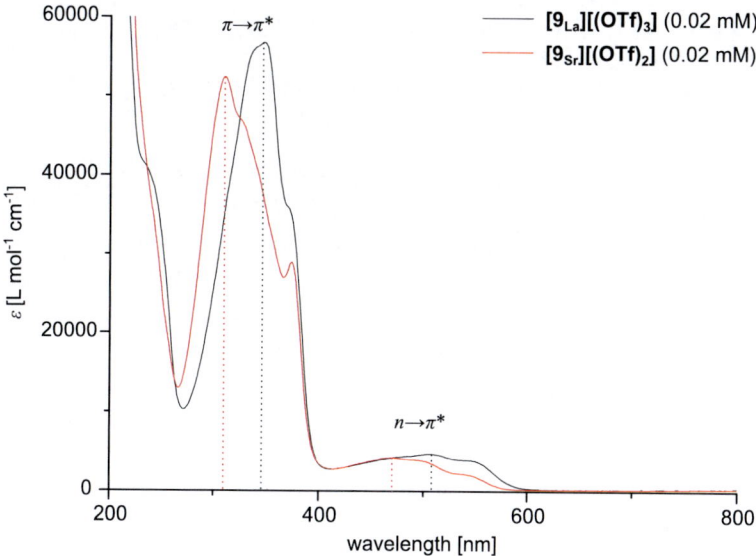

Figure B.2.10. Excerpt of the UV-Vis spectrum of **[9$_{Sr}$][(OTf)$_2$]** and **[9$_{La}$][(OTf)$_3$]** (0.02 mM) in MeCN/MeOH (1:1).

To further elucidate the influence of the anion on the cyclization reaction, CaX$_2$ (X = Cl, Br, I) and Ba(OTs)$_2$ were used as templating agents (Scheme B.2.7). Selective complex formation was observed for all selected anions. **[9$_{Ca}$][Cl$_2$]**, **[9$_{Ca}$][Br$_2$]**, **[9$_{Ca}$][I$_2$]** and **[9$_{Ba}$][(OTs)$_2$]** were prepared similar to **[9$_{Ca}$][(OTf)$_2$]** and obtained in high yield.

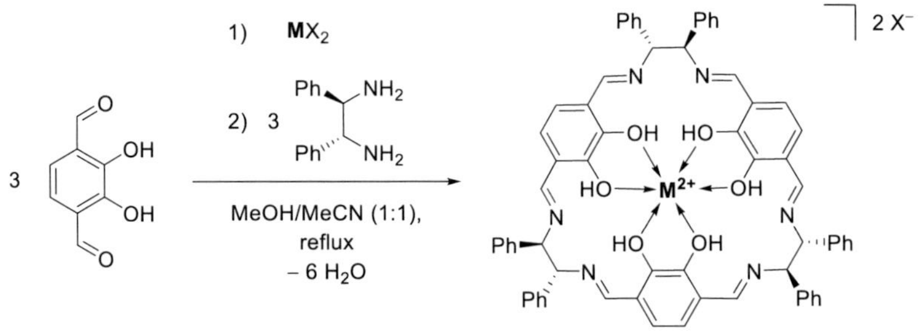

Mononuclear Complexes Featuring a Tris(ONNO)-Type Ligand

Scheme B.2.7. Synthesis of the mononuclear complexes [9$_{Ca}$][Cl$_2$], [9$_{Ca}$][Br$_2$], [9$_{Ca}$][I$_2$] and [9$_{Ba}$][(OTs)$_2$].

The FAB mass spectra of the complexes show prominent peaks at m/z = 1100.3 ([9$_{Ca}$][Cl$_2$]), 1146.3 ([9$_{Ca}$][Br$_2$]), 1192.1 ([9$_{Ca}$][I$_2$]) and 1334.0 ([9$_{Ba}$][(OTs)$_2$]), which correspond to the [M – OTf$^-$]$^+$ fragments. Similar to [9$_{Ca}$][(OTf)$_2$], the ^{1}H NMR spectra of the complexes in MeCN-d_3/MeOD-d_4 (1:1) feature only one set of signals for the imine protons and the aromatic protons of the catechol subunit (Figure B.2.11). A shift of the resonances for the imine protons and the aromatic protons was not observed since the complexes are dissolved as separate ion pairs. As complexes featuring different anions are potentially relevant for application in catalysis, the strontium and barium analogues [9$_{Sr}$][I$_2$] and [9$_{Ba}$][I$_2$] were synthesized similar to [9$_{Ca}$][I$_2$]. The FAB mass spectra show prominent peaks at m/z = 1240.7 [9$_{Sr}$][I$_2$]) and 1290.8 ([9$_{Ba}$][I$_2$]), which correspond to the [M – OTf$^-$]$^+$ fragments. When LaI$_3$ was used as a template, multiple species were observed in the ^{1}H NMR spectrum which may be attributed to unselective complex formation due to very low solubility of LaI$_3$ under the reaction conditions.

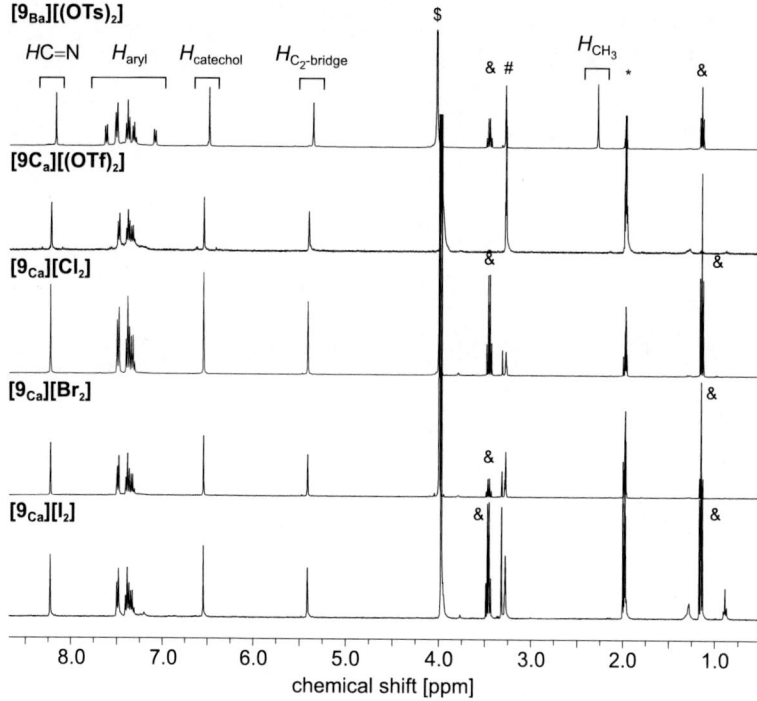

Figure B.2.11. ^{1}H NMR spectra of [9$_{Ba}$][(OTs)$_2$], [9$_{Ca}$][(OTf)$_2$], [9$_{Ca}$][Cl$_2$], [9$_{Ca}$][Br$_2$] and [9$_{Ca}$][I$_2$] in MeCN-d_3/MeOD-d_4 referenced to residual solvent signals (#) MeOD-d_3 and (*) MeCN-d_2 with traces of ($) H$_2$O and (&) Et$_2$O.

Removal of the template cation of mononuclear complexes gives access to protonated pro-ligands. Since the pro-ligands are sensitive toward proton-catalyzed Schiff base rearrangements, the Ca^{2+} template was removed from [6$_{Ca}$][(OTf)$_2$] through precipitation of CaSO$_4$ with an aqueous solution of NaSO$_4$ or guanidinium sulfate to afford the pro-ligand [H$_6^{Cy}$L], **10**, (Scheme B.2.8). Pro-ligand **10** readily produces aggregates which precipitate from solution to produce insoluble residues, complicating separation from the precipitated sulfate salts. Aggregation phenomena are more pronounced on preparative scale reactions, and the structural integrity of the [3+3] macrocycle of aggregated pro-ligands remains indeterminable. **10** is soluble in MeOH and halogenated solvents. The FAB mass spectrum shows a prominent peak at m/z = 733.4 which corresponds to [M + H$^+$]. Retention of the macrocyclic framework is indicated by the absorption of the imine bond at $\tilde{v}$ = 1631 cm^{-1} in the IR spectrum. The IR spectrum does not show the characteristic absorptions of the triflate anion, confirming removal of the template. The imine protons exhibit three singlet resonances at δ 8.19, 8.14 and 8.05 ppm in the ^{1}H NMR spectrum in chloroform-d_1, due to formation of three diastereomers. The aromatic protons exhibit an ABX-type spin system with two doublet resonances at δ 6.66 and

6.58 ppm and a singlet resonance at δ 6.28 ppm. Template removal from the enantiopure complexes **[7$_{Ca}$][(OTf)$_2$]** and **[9$_{Ca}$][(OTf)$_2$]** produces insoluble aggregates with elusive composition.

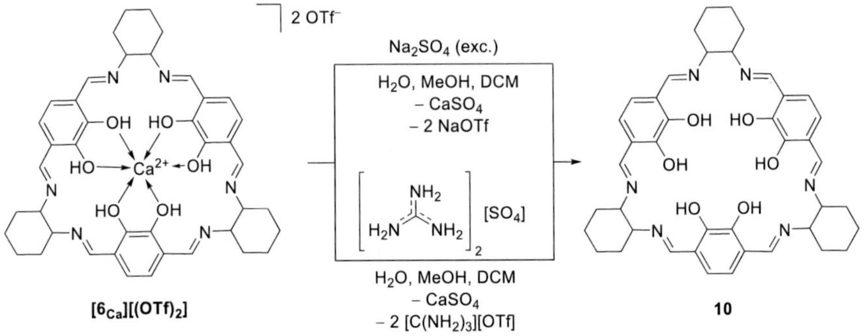

[6$_{Ca}$][(OTf)$_2$]

10

Scheme B.2.8. Removal of the Ca^{2+} template from **[6$_{Ca}$][(OTf)$_2$]** through precipitation of $CaSO_4$ to produce the pro-ligand **10**.

B.2.3 Summary and Outlook

In this chapter, template-directed synthesis of mononuclear complexes featuring [3+3] macrocyclic ligands and vacant salen-type binding sites was reported. These complexes were prepared through pre-coordination of the metal template cations to 3,6-diformylcatechol and subsequent condensation of the dialdehyde with the respective diamines (Scheme B.2.9). Mononuclear cationic alkaline earth metal and lanthanide metal complexes were synthesized, featuring 1,2-cyclohexanediamine or 1,2-diphenylethylenediamine bridging units. Macrocyclic ligands featuring vacant salen-type binding sites, may provide access to heterometallic complexes in a subsequent metalation step.

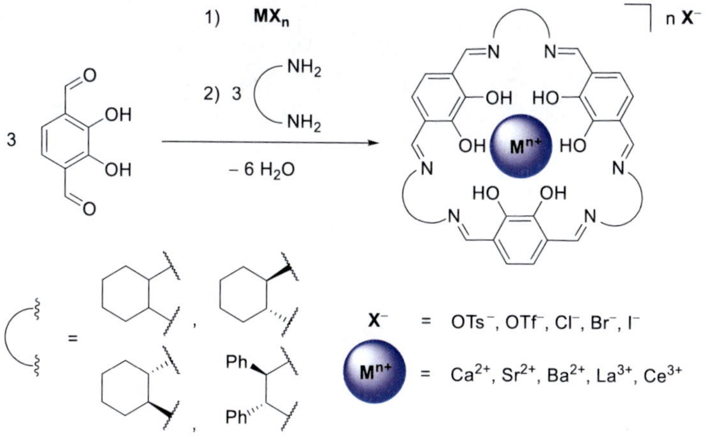

Scheme B.2.9. Template-directed synthesis of mononuclear complexes featuring [3+3] macrocyclic ligands.

Using rac-1,2-cyclohexanediamine, the mononuclear calcium complex $[6_{Ca}][(OTf)_2]$ was synthesized and characterized. Due to the formation of multiple diastereomeric and enantiomeric complexes, the enantiopure diamines (R,R)-1,2-cyclohexanediamine and (S,S)-1,2-cyclohexanediamine were used to prepare enantiopure complexes. The alkaline earth metal cations calcium, strontium and barium, and the lanthanide metal cations lanthanum and cerium(III) are suitable templates to produce the mononuclear complexes $[7_{Ca}][(OTf)_2]$, $[8_{Ca}][(OTf)_2]$, $[7_{Sr}][(OTf)_2]$, $[7_{La}][(OTf)_3]$, $[7_{Ce}][Cl_3]$, $[6_{Ca}][(OTf)_2]$ and $[7_{Ba}][(OTf)_2]$. The compounds were characterized by mass spectrometry, IR and NMR spectroscopy. The template directed synthesis is applicable to several alkaline earth and lanthanide metal cations, enabling preparation of mononuclear complexes featuring coordinated metal cations in the central 18c6-type cavity.

Since complexes featuring the 1,2-cyclohexanediamine bridging unit are poorly soluble in common organic solvents, the complexes [9$_{Ca}$][(OTf)$_2$], [9$_{Sr}$][(OTf)$_2$], [9$_{Ba}$][(OTf)$_2$], [9$_{La}$][(OTf)$_3$] and [9$_{Ce}$][Cl$_3$] featuring a (R,R)-1,2-diphenylethylenediamine bridging unit were synthesized. The complexes were characterized by mass spectrometry, IR and NMR spectroscopy. The molecular structure of [9$_{Ca}$][(OTf)$_2$] was determined by single crystal X-ray diffraction analysis and confirms coordination of the metal cation in the central 18c6-type cavity and retention of the enol-imine tautomer.

To further understand the influence of the templating metal salts onto the cyclization reaction, complexes with different anions were synthesized. The tested tosylate and halide anions did not inhibit the cyclization reaction and the complexes [9$_{Ca}$][Cl$_2$], [9$_{Ca}$][Br$_2$], [9$_{Ca}$][I$_2$], [9$_{Sr}$][I$_2$], [9$_{Ba}$][I$_2$] and [9$_{Ba}$][(OTs)$_2$] were synthesized and characterized similar to [9$_{Ca}$][(OTf)$_2$].

The pro-ligand **10** was synthesized from [6$_{Ca}$][(OTf)$_2$] through template removal with sulfate containing reagents and was characterized by mass spectrometry, IR and NMR spectroscopy. Preparative synthesis of **10** remains challenging, due to formation of insoluble aggregates.

Results and Discussion

B.2.4 Experimental

B.2.4.1 General Considerations

1,4-diformyl-2,3-dimethoxybenzene,[15, 24] 3,6-diformylcatechol, [15, 24] $Sr(OTf)_2$,[25] $Ba(OTf)_2$,[25] $Ba(OTs)_2$[26] and $La(OTf)_3$[27] were prepared according to modified literature procedures. All other chemicals were used without further purification. The monometallic complexes were synthesized under atmospheric conditions without dried solvents.

B.2.4.2 Synthesis of Monometallic Complexes

[$H_6^{Cy}LCa(OTf)_2$] ([6_{Ca}][(OTf)$_2$])

A yellow solution of $Ca(OTf)_2$ (17.0 mg, 50.0 µmol, 1.00 equiv.) and 3,6-diformylcatechol (25.0 mg, 150 µmol, 3.00 equiv.) in MeOH/MeCN (1:1, 10 mL) was heated under refluxing conditions for 15 min to produce a pale-yellow solution. Rac-trans-1,2-diaminocyclohexane (18.9 mg, 170 µmol, 3.30 equiv.) in MeOH/MeCN (1:1, 4 mL) was added to the hot solution upon which the color immediately changed to red. The solution was refluxed for an additional 3 h. All volatiles were removed under reduced pressure to produce a red solid. The solid was rinsed with Et_2O and n-hexane. Drying under reduced pressure afforded **[6_{Ca}][(OTf)$_2$]** as a red solid (80.8 mg, 75.5 µmol, >99%). ^{1}H NMR (DMSO-d_6): δ 13.13 (br. m, 6H, OH), 8.42 (s, 2H, H_aC=N), 8.39 (s, 2H, H_bC=N), 8.38 (s, 2H, H_cC=N) 6.75 (m, 6H, H_{aryl}), 3.42 (m, 6H, HC–N), 1.97 – 1.39 (br. m, 24H, CH_2). ^{13}C{^{1}H} NMR (DMSO-d_6): δ 166.6 (HC_a=N), 165.4 (HC_b=N), 122.3 (C_{aryl}) 120.4 ($C_{a, aryl}$), 120.0 ($C_{b, aryl}$), 118.8 (C_{aryl}), 72.2 (HC–N), 32.6 (CH_2), 32.5 (CH_2), 32.1 (CH_2), 23.8 (CH_2). ^{19}F{^{1}H} NMR (DMSO-d_6): δ –77.7 (s, $F_3CSO_3^-$). **FAB MS** (pos. mode, 3-NBA matrix): m/z = 921.3 ([$H_6^{Cy}LCa(OTf)$]$^+$), 771.3 ([$H_5^{Cy}LCa$]$^+$), 733.4 ([$H_6^{Cy}L$ + H]$^+$). **IR** in KBr: ($\tilde{v}$, cm^{-1}) 3450 (m), 2938 (m), 2863 (m), 1632 (vs, C=N), 1542 (s), 1510 (s), 1459 (s), 1381 (s), 1350 (s), 1276 (vs), 1163 (s), 1104 (w), 1031 (vs), 859 (w), 833 (m), 714 (w), 639 (s), 575 (w), 518 (w).

(R,R)-[$H_6^{Cy}LCa(OTf)_2$] ([7_{Ca}][(OTf)$_2$])

[7_{Ca}][(OTf)$_2$] was synthesized analogously to **[6_{Ca}][(OTf)$_2$]** using the enantiopure (R,R)-(–)-trans-1,2-diaminocyclohexane (69.2 mg, 602 µmol, 3.00 equiv.), 3,6-diformylcatechol (100 mg, 602 µmol, 3.00 equiv.) and $Ca(OTf)_2$ (67.9 mg, 201 µmol, 1.00 equiv.) in MeOH/MeCN (1:1, 50 mL). **[7_{Ca}][(OTf)$_2$]** was isolated as a red solid (171 mg, 160 µmol, 79%). ^{1}H NMR (MeOD-d_4/MeCN-d_3, 1:1): δ 8.51 (s, 6H, HC=N), 6.75 (s, 2H, H_{aryl}), 3.62 (m, 6H, HC–N), 2.49 (br. m, 6H, CH_2), 2.04 (br. m, 6H, CH_2), 1.66 (br. m, 6H, CH_2), 1.47 (br. m, 6H, CH_2). ^{1}H NMR (DMSO-d_6): δ 13.55 – 11.49 (br. m, 6H, OH), 8.44 (s, 6H, HC=N), 6.74 (s, 6H, H_{aryl}), 3.49 (m, 6H, HC–N), 1.97 – 1.39 (br. m, 24H, CH_2). ^{13}C{^{1}H} NMR (DMSO-d_6): δ 161.8 (HC=N), 122.3 (C_{aryl}) 119.1 (C_{aryl}), 117.5 (HC_{aryl}), 70.4 (HC–N), 32.3 (CH_2), 30.8 (CH_2), 24.0 (CH_2), 22.1 (CH_2). ^{19}F{^{1}H}

Mononuclear Complexes Featuring a Tris(ONNO)-Type Ligand

NMR (DMSO-d_6): δ −77.8 (s, $F_3CSO_3^-$). **FAB MS** (pos. mode, 3-NBA matrix): m/z = 921.3 ([H_6^{Cy}LCa(OTf)]$^+$), 771.3 ([H_5^{Cy}LCa]$^+$), 733.4 ([H_6^{Cy}L + H]$^+$). **IR** in KBr: ($\tilde{v}$, cm^{-1}) 3448 (m), 2939 (m), 2864 (m), 1637 (vs, C=N), 1542 (s), 1510 (s), 1459 (s), 1381 (s), 1350 (s), 1276 (vs, sh), 1166 (s), 1106 (w), 1031 (vs), 858 (w), 834 (m), 714 (w), 639 (s), 576 (w), 518 (w).

(S,S)-[H_6^{Cy}LCa(OTf)$_2$] ([8$_{Ca}$][(OTf)$_2$])

[8$_{Ca}$][(OTf)$_2$] was synthesized analogously to [6$_{Ca}$][(OTf)$_2$] using the enantiopure (S,S)-(+)-trans-1,2-diaminocyclohexane. [8$_{Ca}$][(OTf)$_2$] was isolated as a red solid (56.7 mg, 53.0 μmol, >99%). ^{1}**H NMR** (DMSO-d_6): δ 13.55 − 11.49 (br. m, 6H, OH), 8.43 (s, 6H, HC=N), 6.76 (s, 6H, H_{aryl}), 3.47 (m, 6H, HC−N), 1.97 − 1.39 (br. m, 24H, CH$_2$). **IR** in KBr: ($\tilde{v}$, cm^{-1}) 3448 (m), 2939 (m), 2864 (m), 1635 (vs, C=N), 1542 (s), 1509 (s), 1458 (s), 1350 (s), 1276 (vs, sh), 1165(s), 1031 (vs), 860 (w), 833 (m), 703 (w), 639 (s), 576 (w), 518 (w).

(R,R)-[H_6^{Cy}LSr(OTf)$_2$] ([7$_{Sr}$][(OTf)$_2$])

[7$_{Sr}$][(OTf)$_2$] was synthesized analogously to [7$_{Ca}$][(OTf)$_2$] using the enantiopure (R,R)-(−)-trans-1,2-diaminocyclohexane (17.3 mg, 150 μmol, 3.00 equiv.), 3,6-diformylcatechol (25 mg, 150 μmol, 3.00 equiv.) and Sr(OTf)$_2$ (19.3 mg, 50.0 μmol, 1.00 equiv.) in MeOH/MeCN (1:1, 14 mL). [7$_{Sr}$][(OTf)$_2$] was isolated as a red solid (63.1 mg, 44.7 μmol, 89%). ^{1}**H NMR** (MeOD-d_4/MeCN-d_3, 1:1): δ 8.51 (s, 6H, HC=N), 6.76 (s, 6H, H_{aryl}), 3.60 (m, 6H, HC−N), 2.40 (br. m, 6H, CH$_2$), 2.04 (br. m, 6H, CH$_2$), 1.73 (br. m, 6H, CH$_2$), 1.48 (br. m, 6H, CH$_2$).). ^{13}C{^{1}H} **NMR** (MeOD-d_4/MeCN-d_3, 1:1): δ 164.1 (HC=N), 124.1 (C_{aryl}), 118.9 (HC$_{aryl}$), 63.6 (HC−N), 27.2 (CH$_2$), 24.5 (CH$_2$). ^{19}F{^{1}H} **NMR** (MeOD-d_4/MeCN-d_3, 1:1): δ −78.3 (s, $F_3CSO_3^-$). ^{1}**H NMR** (DMSO-d_6): δ 8.49 (s, 6H, HC=N), 6.76 (s, 6H, H_{aryl}), 3.53 (m, 6H, HC−N), 1.97 − 1.39 (br. m, 24H, CH$_2$). **FAB MS** (pos. mode, 3-NBA matrix): m/z = 968.3 ([H_6^{Cy}LSr(OTf) − H]$^+$), 818.5 ([H_5^{Cy}LSr − H]$^+$). **IR** in KBr: ($\tilde{v}$, cm^{-1}) 3341 (w), 3055 (w), 2939 (m), 2862 (m), 1637 (vs, C=N), 1542 (m), 1384 (m), 1350 (m), 1299 (vs), 1273 (vs), 1226 (vs), 1158 (s), 1107 (w), 1030 (s), 914 (w), 903 (w), 850 (w), 835 (m), 760 (w), 720 (w), 638 (s), 605 (w), 573 (w), 517 (w), 446 (w).

(R,R)-[H_6^{Cy}LBa(OTf)$_2$] ([7$_{Ba}$][(OTf)$_2$])

[7$_{Ba}$][(OTf)$_2$] was synthesized analogously to [7$_{Ca}$][(OTf)$_2$] using the enantiopure (R,R)-(−)-trans-1,2-diaminocyclohexane (17.2 mg, 150 μmol, 3.00 equiv.), 3,6-diformylcatechol (25.0 mg, 150 μmol, 3.00 equiv.) and Ba(OTf)$_2$ (21.8 mg, 50.0 μmol, 1.00 equiv.) in MeOH/MeCN (1:1, 14 mL). [7$_{Ba}$][(OTf)$_2$] was isolated as a red solid and purified through

Results and Discussion

precipitation with Et$_2$O/hexane (52.4 mg, 44.9 μmol, 89%). **^{1}H NMR** (MeOD-d_4/MeCN-d_3, 1:1): δ 8.47 (s, 6H, HC=N), 6.68 (s, 6H, H$_{aryl}$), 3.56 (m, 6H, HC–N), 2.42 (br. m, 6H, CH$_2$), 2.03 (br. m, 6H, CH$_2$), 1.65 (br. m, 6H, CH$_2$), 1.48 (br. m, 6H, CH$_2$). **^{13}C{^{1}H} NMR** (MeOD-d_4/MeCN-d_3, 1:1): δ 164.2 (HC=N), 118.9 (HC$_{aryl}$), 116.4 (C$_{aryl}$), 63.5 (HC–N), 28.9 (CH$_2$), 24.7 (CH$_2$). **^{19}F{^{1}H} NMR** (MeOD-d_4/MeCN-d_3, 1:1): δ –78.3 (s, F$_3$CSO$_3^-$). **FAB MS** (pos. mode, 3-NBA matrix): m/z = 1018.4 ([H$_6^{Cy}$LSr(OTf) - H]$^+$), 868.5 ([H$_5^{Cy}$LSr - H]$^+$).

(R,R)-[H$_6^{Cy}$LLa(OTf)$_3$] ([7$_{La}$][(OTf)$_3$])

[7$_{La}$][(OTf)$_3$] was synthesized analogously to [7$_{Ca}$][(OTf)$_2$] using the enantiopure (R,R)-(–)-trans-1,2-diaminocyclohexane (17.3 mg, 150 μmol, 3.00 equiv.), 3,6-diformylcatechol (25.0 mg, 150 μmol, 3.00 equiv.) and La(OTf)$_3$ (29.4 mg, 50.0 μmol, 1.00 equiv.) in MeOH/MeCN (1:1, 14 mL). [7$_{La}$][(OTf)$_3$] was isolated as a red solid (76.5 mg, 47.4 μmol, 95%). **^{1}H NMR** (MeOD-d_4/MeCN-d_3, 1:1): δ 8.53 (s, 6H, HC=N), 6.64 (s, 6H, H$_{aryl}$), 3.75 (m, 6H, HC–N), 2.39 (br. m, 6H, CH$_2$), 2.10 (br. m, 6H, CH$_2$), 1.87 (br. m, 6H, CH$_2$), 1.52 (br. m, 6H, CH$_2$).). **^{13}C{^{1}H} NMR** (MeOD-d_4/MeCN-d_3, 1:1): 164.6 (HC=N), 117.6 (HC$_{aryl}$), 113.9 (C$_{aryl}$), 62.9 (HC–N), 27.9 (CH$_2$), 24.5 (CH$_2$). **^{19}F{^{1}H} NMR** (MeOD-d_4/MeCN-d_3, 1:1): δ –78.3 (s, F$_3$CSO$_3^-$). **FAB MS** (pos. mode, 3-NBA matrix): m/z = 868.3 ([H$_4^{Cy}$LLa]$^+$), 1018.1 ([H$_5^{Cy}$LLa(OTf)]$^+$). **IR** in KBr: ($\tilde{v}$, cm^{-1}) 3442 (m), 2940 (m), 2866 (w), 1634 (vs, C=N), 1585 (m), 1505 (m), 1459 (m), 1384 (s), 1279 (s), 1242 (s), 1165 (s), 1029 (s), 914 (vw), 826 (w), 760 (w), 709 (w), 638 (m), 575 (w), 517 (w), 455 (w).

(R,R)-[H$_6^{Cy}$LCeCl$_3$] ([7$_{Ce}$][Cl$_3$])

[7$_{Ce}$][Cl$_3$] was synthesized analogously to [7$_{Ca}$][(OTf)$_2$] using the enantiopure (R,R)-(–)-trans-1,2-diaminocyclohexane (68.7 mg, 602 μmol, 3.00 equiv.), 3,6-diformylcatechol (100 mg, 602 μmol, 3.00 equiv.) and CeCl$_3$ (49.5 mg, 201 μmol, 1.00 equiv.) in MeOH/MeCN (1:1, 30 mL). [7$_{Ce}$][Cl$_3$] was isolated as a red solid (192 mg, 197 μmol, 98%). **^{1}H NMR** (MeOD-d_4/MeCN-d_3, 1:1): δ 10.51 (s, 6H, HC=N), 8.31 (s, 6H, H$_{aryl}$), 5.24 (m, 6H, HC–N), 3.28 (br. m, 6H, CH$_2$), 2.92 (br. m, 6H, CH$_2$), 2.59 (br. m, 6H, CH$_2$), 2.13 (br. m, 6H, CH$_2$). **^{13}C{^{1}H} NMR** (MeOD-d_4/MeCN-d_3, 1:1): δ 167.7 (HC=N), 124.1 (C$_{aryl}$) 120.7 (HC$_{aryl}$), 118.5 (C$_{aryl}$), 64.7 (HC–N), 29.0 (CH$_2$), 25.6 (CH$_2$). **FAB MS** (pos. mode, 3-NBA matrix): m/z = 870.3 ([H$_4^{Cy}$LCe]$^+$), 907.2 ([H$_5^{Cy}$LCeCl]$^+$). **IR** in KBr: ($\tilde{v}$, cm^{-1}) 3887 (m), 2935 (m), 2861 (m), 1633 (vs), 1506 (m), 1456 (m), 1350 (m), 1305 (m), 1228 (m), 1188 (m), 1107 (w), 1026 (w), 914 (vw), 861 8(vw), 829 (vw), 767 (vw), 707 (vw), 661 (vw), 607 (vw), 562 (vw), 460 (vw). **Elemental Analysis:** Calcd. for C$_{42}$H$_{48}$CeCl$_3$N$_6$O$_6$(H$_2$O)$_8$: C 44.90, H 5.74, N: 7.48; found: C 44.36, H 5.37, N 7.38.

Mononuclear Complexes Featuring a Tris(ONNO)-Type Ligand

(R,R)-[H_6^{C2Ph2}LCa(OTf)$_2$] ([9_{Ca}][(OTf)$_2$])

[9_{Ca}][(OTf)$_2$] was synthesized analogously to [7_{Ca}][(OTf)$_2$] using the enantiopure (R,R)-(+)-1,2-diphenylethylenediamine (31.9 mg, 150 μmol, 3.00 equiv.), 3,6-diformylcatechol (25.0 mg, 150 μmol, 3.00 equiv.) and Ca(OTf)$_2$ (17.0 mg, 50.0 μmol, 1.00 equiv.). [9_{Ca}][(OTf)$_2$] was isolated as a red solid (35.2 mg, 33.0 μmol, 66%). ^{1}H NMR (MeOD-d_4/MeCN-d_3, 1:1): δ 8.22 (s, 6H, HC=N), 7.50 (d, $^3J_{HH}$ = 7.2 Hz, 12H, H_{aryl}), 7.39 (t, $^3J_{HH}$ = 7.1 Hz, 12H, H_{aryl}), 7.33 (t, $^3J_{HH}$ = 7.1 Hz, 6H, H_{aryl}), 6.54 (s, 6H, H_{aryl}), 5.41 (s, 6H, CH). ^{1}H NMR (DMSO-d_6): δ 12.12 (br. m, 6H, OH), 8.59 (s, 6H, HC=N), 7.32 (d, $^3J_{HH}$ = 7.2 Hz, 12H, H_{aryl}), 7.22 (t, $^3J_{HH}$ = 7.1 Hz, 12H, H_{aryl}), 7.15 (t, $^3J_{HH}$ = 7.1 Hz, 6H, H_{aryl}), 6.95 (s, 6H, H_{aryl}), 5.11 (s, 6H, CH). ^{13}C{^{1}H} NMR (MeOD-d_4/MeCN-d_3, 1:1): δ 168.2 (s, HC=N), 163.6 (s, HOC$_{catechol}$), 140.0 (s, C_{aryl}), 130.1 (s, HC$_{aryl}$), 129.6 (s, HC$_{aryl}$), 127.7 (s, HC$_{aryl}$), 119.1 (s, HC$_{aryl,catechol}$), 116.5 (s, $C_{catechol}$), 74.3 (CH$_{C2-bridge}$). ^{19}F{^{1}H} NMR (MeOD-d_4/MeCN-d_3, 1:1): δ −78.3 (s, F_3CSO$_3^−$). **FAB MS** (pos. mode, 3-NBA matrix): m/z = 1215.6 ([H_6^{C2Ph2}LCa(OTf)]$^+$), 1066.6 ([H_6^{C2Ph2}LCa + e$^−$]$^+$). **IR** in KBr: ($\tilde{v}$, cm^{-1}) 3445 (br. m), 3063 (vw), 3033 (vw), 2963 (vw), 2920 (w), 1632 (vs, C=N), 2585 (w), 1543 (m), 1497 (w), 1453 (m), 1362 (m), 1258 (s), 1226 (m), 1169 (m), 1031 (s), 883 (vw), 813 (vw), 774 (vw), 722 (w), 699 (m), 638 (s), 571 (w), 518 (w). **Elemental Analysis:** Calcd. for C$_{68}$H$_{54}$CaF$_6$N$_6$O$_{12}$S$_2$(H$_3$COH)$_2$: C 58.82, H 4.37, N 5.88; found: C 59.00, H 4.19, N 5.98. **XRD:** Crystals suitable for single crystal X-ray diffraction analysis were grown from MeOH/MeCN (1:1) solution layered with Et$_2$O and n-hexane overnight.

(R,R)-[H_6^{C2Ph2}LCaCl$_2$] ([9_{Ca}][Cl$_2$])

[9_{Ca}][Cl$_2$] was synthesized analogously to [9_{Ca}][(OTf)$_2$] using the enantiopure (R,R)-(+)-1,2-diphenylethylenediamine (31.9 mg, 150 μmol, 3.00 equiv.), 3,6-diformylcatechol (25.0 mg, 150 μmol, 3.00 equiv.) and CaCl$_2$ · 2 H$_2$O (7.4 mg, 50.0 μmol, 1.00 equiv.). [9_{Ca}][Cl$_2$] was isolated as a red solid (55.9 mg, 49.0 μmol, 98%). ^{1}H NMR (MeOD-d_4/MeCN-d_3, 1:1): δ 8.22 (s, 6H, HC=N), 7.49 (d, $^3J_{HH}$ = 7 Hz, 12H, H_{aryl}), 7.39 (t, $^3J_{HH}$ = 7 Hz, 12H, H_{aryl}), 7.35 (t, $^3J_{HH}$ = 7 Hz, 6H, H_{aryl}), 6.54 (s, 6H, H_{aryl}), 5.41 (s, 6H, CH). ^{13}C{^{1}H} NMR (MeOD-d_4/MeCN-d_3, 1:1): δ 168.3 (s, HC=N), 163.6 (s, HOC$_{catechol}$), 139.0 (s, C_{aryl}), 130.1 (s, HC$_{aryl}$), 129.6 (s, HC$_{aryl}$), 127.7 (s, HC$_{aryl}$), 119.1 (s, HC$_{aryl,catechol}$), 116.5 (s, $C_{catechol}$), 74.2 (CH$_{C2-bridge}$). **FAB MS** (pos. mode, 3-NBA matrix): m/z = 1100.3 ([H_6^{C2Ph2}LCaCl]$^+$), 1065.4 ([H_5^{C2Ph2}LCa]$^+$), 1218.3 ([H_5^{C2Ph2}LCa + 3-NBA]$^+$). **IR** in KBr: ($\tilde{v}$, cm^{-1}) 3405 (br. m), 3057 (m), 3029 (m), 2970 (m), 2879 (m), 1628 (vs, C=N), 1540 (m), 1495 (m), 1451 (m), 1384 (s), 1362 (m), 1289 (m), 1223 (m), 1181 (m), 1117 (w), 1053 (w), 1028 (w), 1001 (w), 919 (vw), 883 (vw), 812 (vw), 767 (w), 722 (m), 699 (m), 654 (vw), 618 (vw), 568 (vw), 500 (vw), 482 (vw). **Elemental Analysis:** Calcd. for C$_{66}$H$_{54}$CaCl$_2$N$_6$O$_6$(H$_2$O)$_5$: C 64.54, H 5.25, N 6.84; found: C 64.5, H 5.07, N 6.68.

(R,R)-[H₆^{C2Ph2}LCaBr₂] ([9$_{Ca}$][Br₂])

[9$_{Ca}$][Br₂] was synthesized analogously to [9$_{Ca}$][(OTf)₂] using the enantiopure (R,R)-(+)-1,2-diphenylethylenediamine (31.9 mg, 150 μmol, 3.00 equiv.), 3,6-diformylcatechol (25.0 mg, 150 μmol, 3.00 equiv.) and CaBr₂ · 2 H₂O (11.8 mg, 50.0 μmol, 1.00 equiv.). [9$_{Ca}$][Br₂] was isolated as a red solid (51.0 mg, 42.0 μmol, 83%). **¹H NMR** (MeOD-d_4/MeCN-d_3, 1:1): δ 8.23 (s, 6H, HC=N), 7.49 (d, $^3J_{HH}$ = 7 Hz, 12H, H_{aryl}), 7.39 (t, $^3J_{HH}$ = 7 Hz, 12H, H_{aryl}), 7.35 (t, $^3J_{HH}$ = 7 Hz, 6H, H_{aryl}), 6.54 (s, 6H, H_{aryl}), 5.41 (s, 6H, CH). **¹³C{¹H} NMR** (MeOD-d_4/MeCN-d_3, 1:1): δ 168.3 (s, HC=N), 163.6 (s, HO$C_{catechol}$), 139.0 (s, C_{aryl}), 130.1 (s, HC_{aryl}), 129.6 (s, HC_{aryl}), 127.7 (s, HC_{aryl}), 119.1 (s, H$C_{aryl,catechol}$), 116.5 (s, $C_{catechol}$), 74.2 (CH$_{C2\text{-bridge}}$). **FAB MS** (pos. mode, 3-NBA matrix): m/z = 1065.4 ([H₆^{C2Ph2}LCa + 3-NBA]⁺), 1146.3 ([H₆^{C2Ph2}LCaBr]⁺). **IR** in KBr: ($\tilde{v}$, cm⁻¹) 3405 (m), 3059 (w), 3029 (w), 2884 (w), 1629 (vs, C=N), 1542 (m), 1496 (m), 1451 (m), 1384 (m), 1357 (m), 1318 (m), 1289 (m), 1224 (m), 1184 (w), 1158 (vw), 1117 (vw), 1059 (w), 1029 (w), 1001 (w), 927 (vw), 879 (vw), 816 (vw), 756 (vw), 724 (m), 701 (m), 655 (vw), 569 (vw), 501 (vw), 482 (vw). **Elemental Analysis:** Calcd. for C₆₆H₅₄CaBr₂N₆O₆(H₂O)₃: C 61.88, H 4.72, N 6.56; found: C 61.61, H 4.68, N 6.42.

(R,R)-[H₆^{C2Ph2}LCaI₂] ([9$_{Ca}$][I₂])

[9$_{Ca}$][I₂] was synthesized analogously to [9$_{Ca}$][(OTf)₂] using the enantiopure (R,R)-(+)-1,2-diphenylethylenediamine (128 mg, 602 μmol, 3.00 equiv.), 3,6-diformylcatechol (100 mg, 602 μmol, 3.00 equiv.) and CaI₂ (59.0 mg, 201 μmol, 1.00 equiv.). [9$_{Ca}$][I₂] was isolated as a red solid (265 mg, 201 μmol, >99%). **¹H NMR** (MeOD-d_4/MeCN-d_3, 1:1): δ 8.23 (s, 6H, HC=N), 7.49 (d, $^3J_{HH}$ = 7 Hz, 12H, H_{aryl}), 7.39 (t, $^3J_{HH}$ = 7 Hz, 12H, H_{aryl}), 7.35 (t, $^3J_{HH}$ = 7 Hz, 6H, H_{aryl}), 6.54 (s, 6H, H_{aryl}), 5.41 (s, 6H, CH).). **FAB MS** (pos. mode, 3-NBA matrix): m/z = 1192.1 ([H₆^{C2Ph2}LCaI - H]⁺), 1065.3 ([H₅^{C2Ph2}LCa]⁺). **IR** in KBr: ($\tilde{v}$, cm⁻¹) 3407 (m), 3029 (w), 2918 (w), 1630 (vs, C=N), 1542 (m), 1496 (m), 1452 (m), 1384 (m), 1364 (m), 1289 (m), 1226 (m), 1183 (w), 1158 (w), 1114 (vw), 1089 (vw), 1053 (w), 1029 (w), 1002 (w), 813 (w), 766 (w), 722 (w), 699 (m), 570 (w). **Elemental Analysis:** Calcd. for C₆₆H₅₄CaI₂N₆O₆: C 60.01, H 4.12, N 6.36; found: C 57.29, H 9.04, N 6.20.

(R,R)-[H₆^{C2Ph2}LSr(OTf)₂] ([9$_{Sr}$][(OTf)₂])

[9$_{Sr}$][(OTf)₂] was synthesized analogously to [9$_{Ca}$][(OTf)₂] using the enantiopure (R,R)-(+)-1,2-diphenylethylenediamine (31.9 mg, 150 μmol, 3.00 equiv.), 3,6-diformylcatechol (25.0 mg, 150 μmol, 3.00 equiv.) and Sr(OTf)₂ (19.3 mg, 50 μmol, 1.00 equiv.). [9$_{Sr}$][(OTf)₂] was isolated as a red solid (135 mg, 95.7 μmol, 96%). **¹H NMR** (MeOD-d_4/MeCN-d_3, 1:1): δ 8.25 (s, 6H, HC=N), 7.50 (d, $^3J_{HH}$ = 7 Hz, 12H, H_{aryl}), 7.39 (t, $^3J_{HH}$ = 7 Hz, 12H, H_{aryl}), 7.35 (t, $^3J_{HH}$ = 7 Hz, 6H, H_{aryl}), 6.55 (s, 6H, H_{aryl}), 5.40 (s, 6H, CH). **¹³C{¹H} NMR** (MeOD-d_4/MeCN-d_3, 1:1): δ 168.4

(s, HC=N), 163.7 (s, HO$C_{catechol}$), 139.6 (s, C_{aryl}), 130.1 (s, HC_{aryl}), 129.5 (s, HC_{aryl}), 127.8 (s, HC_{aryl}), 119.2 (s, H$C_{aryl,catechol}$), 117.0 (s, $C_{catechol}$), 74.0 (C$H_{C2\text{-bridge}}$). ^{19}F{^{1}H} NMR (MeOD-d_4/MeCN-d_3, 1:1): δ −78.3 (s, F_3CSO$_3^-$). FAB MS (pos. mode, 3-NBA matrix): m/z = 1262.1 ([H$_6$^{C2Ph2}LSr(OTf) − H]$^+$), 1113.2 ([H$_5$^{C2Ph2}LSr]$^+$). IR in KBr: ($\tilde{\nu}$, cm^{-1}) 3396 (br. w), 3030 (w), 2973 (w), 2932 (w), 2894 (w), 1631 (vs), 1585 (m), 1545 (s), 1498 (s), 1451 (s), 1417 (m), 1384 (s), 1364 (s), 1257 (vs), 1223 (s), 1171 (s), 1088 (m), 1053 (m), 1026 (s), 1002 (m), 932 (s), 912 (w), 883 (w), 823 (w), 812 (w), 774 (w), 720 (m), 695 (m), 637 (s), 573 (m), 516 (m), 480 (w). **Elemental Analysis** Calcd. for C$_{68}$H$_{54}$SrF$_6$N$_6$O$_{12}$S$_2$: C 57.80, H 3.85, N 5.95; found: C 54.55, H 6.45, N 4.15.

(R,R)-[H$_6$^{C2Ph2}LSrI$_2$] ([9$_{Sr}$][I$_2$])

[9$_{Sr}$][I$_2$] was synthesized analogously to [9$_{Ca}$][(OTf)$_2$]] using the enantiopure (R,R)-(+)-1,2-diphenylethylenediamine (95.8 mg, 451 µmol, 3.00 equiv.), 3,6-diformylcatechol (75.0 mg, 451 µmol, 3.00 equiv.) and SrI$_2$ (51.4 mg, 150 µmol, 1.00 equiv.). [9$_{Sr}$][I$_2$] was isolated as a red solid (187 mg, 137 µmol, 91%). ^{1}H NMR (MeOD-d_4/MeCN-d_3, 1:1): δ 8.25 (s, 6H, HC=N), 7.50 (d, $^3J_{HH}$ = 7 Hz, 12H, H_{aryl}), 7.39 (t, $^3J_{HH}$ = 7 Hz, 12H, H_{aryl}), 7.33 (t, $^3J_{HH}$ = 7 Hz, 6H, H_{aryl}), 6.55 (s, 6H, H_{aryl}), 5.41 (s, 6H, CH).). ^{13}C{^{1}H} NMR (MeOD-d_4/MeCN-d_3, 1:1): δ 168.4 (s, HC=N), 150.8 (s, HO$C_{catechol}$), 139.6 (s, C_{aryl}), 130.1 (s, HC_{aryl}), 129.6 (s, HC_{aryl}), 127.8 (s, HC_{aryl}), 119.2 (s, H$C_{aryl,catechol}$), 117.1 (s, $C_{catechol}$), 74.0 (C$H_{C2\text{-bridge}}$). FAB MS (pos. mode, 3-NBA matrix): m/z = 1240.7 ([H$_6$^{C2Ph2}LSrI]$^+$), 1112.8 ([H$_5$^{C2Ph2}LSr]$^+$). IR in KBr: ($\tilde{\nu}$, cm^{-1}) 3418 (br. w), 3060 (w), 2973 (w), 3031 (w), 2989 (w), 2887 (w), 1623 (vs, C=N), 1582 (m), 1542 (m), 1496 (m), 1450 (m), 1384 (m), 1355 (m), 1320 (m), 1288 (m), 1223 (m), 1185 (w), 1159 (vw), 1081 (vw), 1063 (w), 1029 (w), 999 (w), 948 (vw), 887 (vw), 819 (vw), 782 (w), 764 (w), 726 (s), 700 (m), 655 (vw), 617 (vw), 609 (vw), 570 (vw), 500 (vw), 482 (vw). **Elemental Analysis:** Calcd. for C$_{66}$H$_{54}$SrI$_2$N$_6$O$_6$(H$_2$O): C 57.17, H 4.07, N 6.06; found: C 57.06, H 4.04, N 5.91.

(R,R)-[H$_6$^{C2Ph2}LBa(OTf)$_2$] ([9$_{Ba}$][(OTf)$_2$])

[9$_{Ba}$][(OTf)$_2$] was synthesized analogously to [9$_{Ca}$][(OTf)$_2$] using the enantiopure (R,R)-(+)-1,2-diphenylethylenediamine (63.9 mg, 301 µmol, 3.00 equiv.), 3,6-diformylcatechol (50.0 mg, 301 µmol, 3.00 equiv.) and Ba(OTf)$_2$ (43.0 mg, 100 µmol, 1.00 equiv.). [9$_{Ba}$][(OTf)$_2$] was isolated as a red solid (119 mg, 81.4 µmol, 81%). ^{1}H NMR (MeOD-d_4/MeCN-d_3, 1:1): δ 8.18 (s, 6H, HC=N), 7.51 (d, $^3J_{HH}$ = 7 Hz, 12H, H_{aryl}), 7.39 (t, $^3J_{HH}$ = 7 Hz, 12H, H_{aryl}), 7.33 (t, $^3J_{HH}$ = 7 Hz, 6H, H_{aryl}), 6.50 (s, 6H, $H_{aryl,catechol}$), 5.38 (s, 6H, CH). ^{13}C{^{1}H} NMR (MeOD-d_4/MeCN-d_3, 1:1): δ 168.4 (s, HC=N), 164.2 (s, HO$C_{catechol}$), 140.2 (s, C_{aryl}), 130.0 (s, HC_{aryl}), 129.5 (s, HC_{aryl}), 127.9 (s, HC_{aryl}), 119.2 (s, H$C_{aryl,catechol}$), 117.4 (s, $C_{catechol}$), 74.0 (s, H$C_{C2\text{-bridge}}$). ^{19}F{^{1}H} NMR (MeOD-

d_4/MeCN-d_3, 1:1): δ −78.3 (s, $F_3CSO_3^-$). **FAB MS** (pos. mode, 3-NBA matrix): m/z = 1313.3 ([H_6^{C2Ph2}LBa(OTf)]$^+$), 1163.3 ([H_5^{C2Ph2}LBa]$^+$). **IR** in KBr: ($\tilde{\nu}$, cm^{-1}) 3349 (w), 3061 (w), 3032 (w), 2892 (w), 1621 (vs, C=N), 1550 (s), 1497 (s), 1452 (s), 1283 (vs), 1156 (s), 1089 (m), 1059 (m), 1030 (vs), 1001 (s), 918 (w), 812 (w), 777 (w), 750 (w), 722 (s), 700 (s), 639 (vs), 573 (m), 571 (m), 498 (w), 481 (w). **Elemental Analysis:** Calcd. for $C_{68}H_{54}BaF_6N_6O_{12}S_2$: C 55.84, H 3.72, N 5.75; found: C 55.37, H 3.74, N 5.70.

(R,R)-[H_6^{C2Ph2}LBa(OTs)$_2$] ([9_{Ba}][(OTs)$_2$])

[9_{Ba}][(OTs)$_2$] was synthesized analogously to [9_{Ca}][(OTf)$_2$] using the enantiopure (R,R)-(+)-1,2-diphenylethylenediamine (31.9 mg, 150 μmol, 3.00 equiv.), 3,6-diformylcatechol (25.0 mg, 150 μmol, 3.00 equiv.) and Ba(OTs)$_2$ (24.1 mg, 50.0 μmol, 1.00 equiv.). [9_{Ba}][(OTs)$_2$] was isolated as a red solid (70.8 mg, 47.0 μmol, 94%). ^{1}**H NMR** (MeOD-d_4/MeCN-d_3, 1:1): δ 8.16 (s, 6H, HC=N), 7.63 (d, $^3J_{HH}$ = 8 Hz, 4H, $H_{tosylate}$), 7.50 (d, $^3J_{HH}$ = 7 Hz, 12H, H_{aryl}), 7.38 (t, $^3J_{HH}$ = 7 Hz, 12H, H_{aryl}), 7.31 (t, $^3J_{HH}$ = 7 Hz, 6H, H_{aryl}), 7.11 (d, $^3J_{HH}$ = 8 Hz, 4H, $H_{tosylate}$), 6.55 (s, 6H, H_{aryl}), 5.36 (s, 6H, CH), 2.30 (s, 6H, CH_3). ^{13}**C{^{1}H} NMR** (MeOD-d_4/MeCN-d_3, 1:1): δ 168.3 (s, HC=N), 164.0 (s, HO$C_{catechol}$), 143.9 (s, $C_{aryl,tosylate}$), 141.3 (s, $C_{aryl,tosylate}$), 140.2 (s, C_{aryl}), 130.0 (s, HC_{aryl}), 129.7 (s, H$C_{aryl,tosylat}$), 129.4 (s, HC_{aryl}), 128.0 (s, HC_{aryl}), 126.8(s, H$C_{aryl,tosylate}$), 119.1 (s, H$C_{aryl,catechol}$) 117.4 (s, $C_{catechol}$), 74.1 (s, H$C_{C2-bridge}$), 21.3 (s, CH_3). **FAB MS** (pos. mode, 3-NBA matrix): m/z = 1334.0 ([H_6^{C2Ph2}LBa(OTs)]$^+$), 1162.1 ([H_5^{C2Ph2}LBa]$^+$). **IR** in KBr: ($\tilde{\nu}$, cm^{-1}) 3431 (br. m), 3056 (w), 3030 (w), 2920 (w), 1625 (vs, C=N), 1584 (m), 1496 (m), 1384 (s), 1288 (m), 1223 (s), 1188 (s), 1122 (s), 1055 (w), 1033 (m), 1011 (m), 935 (vw), 814 (m), 775 (w), 723 (m), 700 (m), 682 (m), 619 (vw), 570 (m), 501 (vw). **Elemental Analysis:** Calcd. for $C_{80}H_{68}BaN_6O_{12}S_2(H_2O)_2$: C 62.28, H 4.70, N 5.45; found: C 61.90, H 4.45, N 5.37.

(R,R)-[H_6^{C2Ph2}LBaI$_2$] ([9_{Ba}][I$_2$])

[9_{Ba}][I$_2$] was synthesized analogously to [9_{Ca}][(OTf)$_2$] using the enantiopure (R,R)-(+)-1,2-diphenylethylenediamine (95.8 mg, 451 μmol, 3.00 equiv.), 3,6-diformylcatechol (75.0 mg, 451 μmol, 3.00 equiv.) and BaI$_2$ (58.9 mg, 150 μmol, 1.00 equiv.). [9_{Ba}][I$_2$] was isolated as a red solid (204 mg, 144 μmol, 96%). ^{1}**H NMR** (MeOD-d_4/MeCN-d_3, 1:1): δ 8.18 (s, 6H, HC=N), 7.50 (d, $^3J_{HH}$ = 7 Hz, 12H, H_{aryl}), 7.39 (t, $^3J_{HH}$ = 7 Hz, 12H, H_{aryl}), 7.32 (t, $^3J_{HH}$ = 7 Hz, 6H, H_{aryl}), 6.50 (s, 6H, H_{aryl}), 5.38 (s, 6H, CH).). **FAB MS** (pos. mode, 3-NBA matrix): m/z = 1162.9 ([H_5^{C2Ph2}LBa]$^+$), 1290.8 ([H_6^{C2Ph2}LBaI]$^+$). **IR** in KBr: ($\tilde{\nu}$, cm^{-1}) 3424 (m), 3059 (w), 3030 (w), 2988 (w), 2890 (w), 1621 (vs, C=N), 1548 (m), 1495 (m), 1449 (m), 1384 (s), 1354 (m), 1224 (m), 1183 (w), 1159 (w), 1117 (w), 1062 (w), 1028 (w), 999 (w), 941 (vw), 926 (vw), 818 (w), 781 (w), 754 (w), 725 (s), 700 (m), 617 (vw), 572 (w), 498 (w), 482 (w). **Elemental Analysis:** Calcd. for $C_{66}H_{54}BaI_2N_6O_6$: C 55.89, H 3.84, N 5.93; found: C 55.38, H 3.91, N 5.73.

(R,R)-[H$_6$^{C2Ph2}LLa(OTf)$_3$] ([9$_{La}$][(OTf)$_3$])

[9$_{La}$][(OTf)$_3$] was synthesized analogously to **[9$_{Ca}$][(OTf)$_2$]** using the enantiopure (R,R)-(+)-1,2-diphenylethylenediamine (128 mg, 602 μmol, 3.00 equiv.), 3,6-diformylcatechol (100 mg, 602 μmol, 3.00 equiv.) and La(OTf)$_3$ (118 mg, 201 μmol, 1.00 equiv.). **[9$_{La}$][(OTf)$_3$]** was isolated as a red solid (320 mg, 198 μmol, 99%). **^{1}H NMR** (MeOD-d_4/MeCN-d_3, 1:1): δ 8.20 (s, 6H, HC=N), 7.49 (d, $^3J_{HH}$ = 7 Hz, 12H, H_{aryl}), 7.40 (m, 18H, H_{aryl}), 6.41 (s, 6H, H_{aryl}), 5.68 (s, 6H, CH). **^{13}C{^{1}H} NMR** (MeOD-d_4/MeCN-d_3, 1:1): δ 172.1 (s, HOC$_{catechol}$), 168.7 (s, HC=N), 136.4 (s, C_{aryl}), 130.4 (s, HC$_{aryl}$), 130.2 (s, HC$_{aryl}$), 128.0 (s, HC$_{aryl}$), 117.9 (s, HC$_{aryl,catechol}$), 114.7 (s, $C_{catechol}$), 71.6 (s, HC$_{C2-bridge}$). **^{19}F{^{1}H} NMR** (MeOD-d_4/MeCN-d_3, 1:1): δ −78.2 (s, $F_3CSO_3^-$). **FAB MS** (pos. mode, 3-NBA matrix): m/z = 1312.0 ([H$_4$^{C2Ph2}LLa(OTf) − e$^-$]$^+$), 1162.2 ([H$_3$^{C2Ph2}LLa − e$^-$]$^+$). **IR** in KBr: ($\tilde{v}$, cm^{-1}) 3431 (br. w), 3062 (w), 3033 (w), 2973 (w), 2933 (w), 1630 (vs, C=N), 1586 (w), 1548 (m), 1514 (m), 1498 (m), 1454 (m), 1384 (s), 1365 (m), 1270 (s), 1237 (s), 1222 (s), 1168 (s), 1057 (s), 1029 (s), 914 (vw), 888 (vw), 843 (vw), 816 (vw), 760 (w), 722 (w), 698 (m), 637 (s), 572 (w), 516 (w), 484 (vw). **Elemental Analysis:** Calcd. for $C_{69}H_{54}F_9LaN_6O_{15}S_3(H_2O)_2$: C 50.25, H 3.54, N 5.10; found: C 50.17, H 3.71, N 5.11.

(R,R)-[H$_6$^{C2Ph2}LCeCl$_3$] ([9$_{Ce}$][Cl$_3$])

[9$_{Ce}$][Cl$_3$] was synthesized analogously to **[9$_{Ca}$][(OTf)$_2$]** using the enantiopure (R,R)-(+)-1,2-diphenylethylenediamine (128 mg, 602 μmol, 3.00 equiv.), 3,6-diformylcatechol (100 mg, 602 μmol, 3.00 equiv.) and CeCl$_3$ (49.5 mg, 201 μmol, 1.00 equiv.). **[9$_{Ce}$][Cl$_3$]** was isolated as a red solid (234 mg, 184 μmol, 91%). **^{1}H NMR** (MeOD-d_4/MeCN-d_3, 1:1): δ 9.98 (s, 6H, s, 6H, HC=N), 8.38 (d, $^3J_{HH}$ = 5.2 Hz, 12H, H_{aryl}), 7.94 (s, 6H, $H_{aryl,catechol}$), 7.47 − 7.31 (m, 18H, H_{aryl}), 7.02 (s, 6H, CH). **^{13}C{^{1}H} NMR** (MeOD-d_4/MeCN-d_3, 1:1): δ 179.1 (s, COH), 170.9 (s, HC=N), 136.8 (s, C_{aryl}), 130.6 (s, HC$_{aryl}$) 130.4 (s, HC$_{aryl}$), 129.2 (s, HC$_{aryl}$), 123.9 (s, $C_{aryl,catechol}$), 120.6 (s, HC$_{aryl,catechol}$), 73.1 (s, HC$_{C2-bridge}$). **FAB MS** (pos. mode, 3-NBA matrix): m/z = 1200.3 ([H$_5$^{C2Ph2}LCeCl]$^+$), 1164.3 ([H$_4$^{C2Ph2}LCe]$^+$). **IR** in KBr: ($\tilde{v}$, cm^{-1}) 3411 (w), 3029 (w), 2922 (w), 1627 (vs), 1548 (m), 1498 (m), 1452 (m), 1362 (m), 1272 (m), 1222 (m), 1183 (m), 1050 (w), 1029 (w), 887 (vw), 819 (vw), 765 (w), 729 (w), 698 (w), 567 (w). **Elemental Analysis** Anal. Calcd. for $C_{66}H_{54}CeCl_3N_6O_6(H_2O)_6$: C 57.37, H 4.81, N 6.08; found: C 56.78, H 5.08, N 5.24.

B.2.4.3 Synthesis of Metal-Free Pro-Ligand
[$H_6^{Cy}L$] (10)

A saturated aqueous solution of Na_2SO_4 (0.5 mL) was added to a suspension of **[6_{Ca}][(OTf)$_2$]** (10 mg, 9.3 µmol) in DCM (1 mL). MeOH was added as a co-solvent until the red starting compound completely dissolved to produce a homogeneous orange solution and a white precipitate. The solution was collected through filtration and the white solid washed with DCM. All volatiles were removed under reduced pressure from the combined liquid phases. The orange solid was extracted with DCM to remove residual salts. Removal of all volatiles afforded the product **[$H_6^{Cy}L$]** as an orange solid (6.8 mg, 9.3 µmol, >99%). ^{1}**H NMR** (CDCl$_3$): δ 13.87 – 12.88 (br. s, 6H, O*H*), 8.19 (s, 2H, *H$_a$*C=N), 8.14 (s, 2H, *H$_b$*C=N), 8.05 (s, 2H, *H$_c$*C=N), 6.66 (d, 2H, J_{AB} = 8.0 Hz, $H_{A, aryl}$), 6.58 (d, 2H, J_{AB} = 8.5 Hz, $H_{B, aryl}$), 6.28 (s, 2H, $H_{X, aryl}$), 3.29 (m, 6H, *H*C-N), 2.09 – 1.19 (m., 24H, C*H$_2$*). **FAB MS** (pos. mode, 3-NBA matrix): *m*/*z* = 733.4 ([$H_6^{Cy}L$ + H]$^+$). **IR** in KBr: ($\tilde{v}$, cm^{-1}) 3435 (br m), 2931 (m), 2858 (m), 1631 (vs, C=N), 1561 (w), 1431 (m), 1345 (w), 1296 (m), 1230 (w), 1145 (w), 1091 (w), 1031 (w), 862 (vw), 823 (w), 638 (vw), 618 (vw).

B.2.5 References

1. Pedersen, C. J. *J. Am. Chem. Soc.* **1967**, *89*, 7017-7036.
2. Schroeder, H. E.; Pedersen, C. J. *Pure Appl. Chem.* **1988**, *60*, 445-451.
3. Christensen, J. J.; Eatough, D. J.; Izatt, R. M. *Chem. Rev.* **1974**, *74*, 351-384.
4. Bradshaw, J. S.; Stott, P. E. *Tetrahedron* **1980**, *36*, 461-510.
5. Dietrich, B.; Viout, P.; Lehn, J.-M., *Macrocyclic Chemistry: Aspects of Organic and Inorganic Supramolecular Chemistry*. VCH Verlagsgesellschaft: Weinheim, 1993.
6. Lichtenberg, C.; Jochmann, P.; Spaniol, T. P.; Okuda, J. *Angew. Chem. Int. Ed.* **2011**, *50*, 5753-5756.
7. Van Veggel, F. C. J. M.; Verboom, W.; Reinhoudt, D. N. *Chem. Rev.* **1994**, *94*, 279-299.
8. Verboom, W.; Rudkevich, D. M.; Reinhoudt, D. N. *Pure Appl. Chem.* **1994**, *66*, 679-686.
9. M. G. Antonisse, M.; N. Reinhoudt, D. *Chem. Commun.* **1998**, *0*, 443-448.
10. Van Veggel, F. C. J. M.; Harkema, S.; Bos, M.; Verboom, W.; Woolthuis, G. K.; Reinhoudt, D. N. *J. Org. Chem.* **1989**, *54*, 2351-2359.
11. Van Veggel, F. C. J. M.; Harkema, S.; Bos, M.; Verboom, W.; Van Staveren, C. J.; Gerritsma, G. J.; Reinhoudt, D. N. *Inorg. Chem.* **1989**, *28*, 1133-1148.
12. Huck, W. T. S.; van Veggel, F. C. J. M.; Reinhoudt, D. N. *Recl. Trav. Chim. Pays-Bas* **1995**, *114*, 273-276.
13. Gallant, A. J.; Yun, M.; Sauer, M.; Yeung, C. S.; MacLachlan, M. J. *Org. Lett.* **2005**, *7*, 4827-4830.
14. Akine, S.; Utsuno, F.; Piao, S.; Orita, H.; Tsuzuki, S.; Nabeshima, T. *Inorg. Chem.* **2016**, *55*, 810-821.
15. Kleemann, J. Homogenkatalysatoren für die Copolymerisation von Cyclohexenoxid mit CO_2 auf Basis multimetallischer Komplexe von Seltenerdmetallen und Zink. Doctoral Thesis, RWTH Aachen University, Aachen, 2017.
16. Akine, S.; Sunaga, S.; Taniguchi, T.; Miyazaki, H.; Nabeshima, T. *Inorg. Chem.* **2007**, *46*, 2959-2961.
17. Akine, S.; Taniguchi, T.; Dong, W.; Masubuchi, S.; Nabeshima, T. *J. Org. Chem.* **2005**, *70*, 1704-1711.
18. Johnston, D. H.; Shiver, D. F. *Inorg. Chem.* **1993**, *32*, 1045-1047.
19. Shannon, R. D. *Acta Cryst. A* **1976**, *32*, 751-767.
20. Spek, A. L. *J. Appl. Cryst.* **2003**, *36*, 7-13.
21. Steinbauer, J.; Spannenberg, A.; Werner, T. *Green Chem.* **2017**, *19*, 3769-3779.
22. Akine, S.; Taniguchi, T.; Nabeshima, T. *Tetrahedron Lett.* **2001**, *42*, 8861-8864.
23. Gallant, A. J.; Hui, J. K. H.; Zahariev, F. E.; Wang, Y. A.; MacLachlan, M. J. *J. Org. Chem.* **2005**, *70*, 7936-7946.
24. Grudzien, K.; Malinska, M.; Barbasiewicz, M. *Organometallics* **2012**, *31*, 3636-3646.
25. Hall, C. D.; Djedovic, N. *J. Organomet. Chem.* **2002**, *648*, 8-13.
26. Frankland, A. D.; Lappert, M. F. *J. Chem. Soc., Dalton Trans.* **1996**, *0*, 4151-4152.
27. Smith, P. H.; Raymond, K. N. *Inorg. Chem.* **1985**, *24*, 3469-3477.

Results and Discussion

B.3 Heterometallic Tetranuclear Complexes Featuring a Tris(ONNO)-Type Ligand

B.3.1 Introduction

Multinuclear complexes have received increasing attention due to their potential application in chemical reactions, small molecule activation, and their unique magnetic and electronic properties.[1-4] Due to the close proximity of the metal centers, these systems are interesting models for the study of metallic cooperativity, which is frequently encountered in active sites of metalloenzymes.[1] Important natural metalloenzymes that contain co-factors which rely on metal cooperativity are nitrogenases[5] and the photosystem II (PSII).[6] Due to positive cooperative effects, these enzymes catalyze vital biological reactions under ambient reaction conditions. The molybdenum iron nitrogenase utilizes three metallo-cofactors based on iron sulfur clusters [Fe_4S_4] and [Fe_8S_7], as well as a heterometallic iron molybdenum [$MoFe_7S_9C$] cluster to reduce N_2 to bio-available ammonia (Figure B.3.1(a)).[5] The oxygen-evolving complex (OEC) of PSII consists of a tetranuclear [Mn_4CaO_n] cluster which oxidizes water to O_2 (Figure B.3.1(b)).[7]

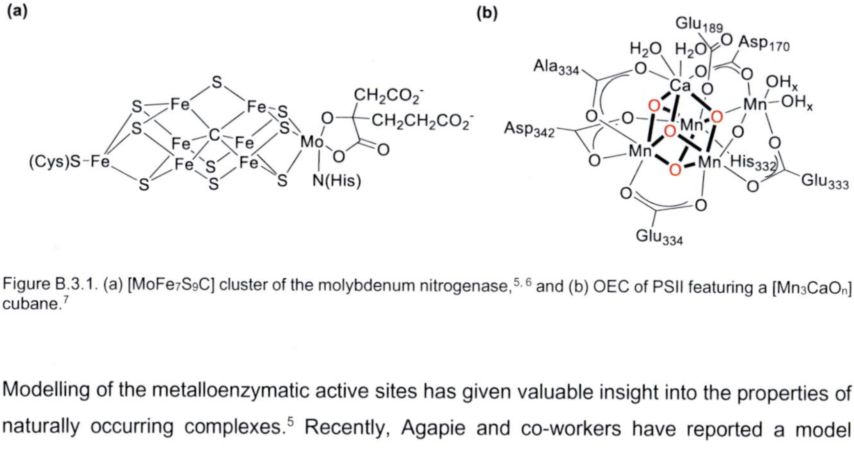

Figure B.3.1. (a) [$MoFe_7S_9C$] cluster of the molybdenum nitrogenase,[5, 6] and (b) OEC of PSII featuring a [Mn_3CaO_n] cubane.[7]

Modelling of the metalloenzymatic active sites has given valuable insight into the properties of naturally occurring complexes.[5] Recently, Agapie and co-workers have reported a model system that most closely resembles the structural motif of the [Mn_3CaO_4] cubane of PSII, albeit without the dangling fourth manganese center (Figure B.3.2).[8-10] The [MMn_3O_4] (M = Mn, Sc, Y, Zn, Ca, Sr) cubane models are structurally related to the naturally occurring cluster and are substituted with redox-inactive metal centers M. Upon variation of the Lewis acidity of the capping metal M, the redox potentials of the [$MMn(IV)_3O_4$]/[$MMn(IV)_2Mn(III)O_4$] couples are shifted up to 1 V. Electrochemical analysis of the structural model complexes thus supports that Ca^{2+} modulates the redox properties of the natural [Mn_4CaO_n] cluster to facilitate electron transfer processes. Ultimately, a deeper understanding of naturally occurring multimetallic complexes may lead to new compounds which utilize metallic cooperativity in catalytic reactions.

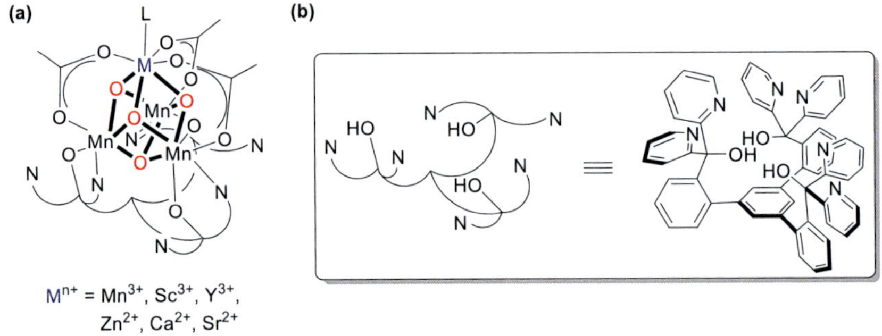

Heterometallic Tetranuclear Complexes Featuring a Tris(ONNO)-Type Ligand

(a)

M^{n+} = Mn^{3+}, Sc^{3+}, Y^{3+}, Zn^{2+}, Ca^{2+}, Sr^{2+}

(b)

Figure B.3.2. (a) Model complex of the [Mn_3CaO_4] cubane of PSII by Agapie and co-workers (L = coordinating solvent) and (b) stabilizing ligand.[8, 9]

Due to the versatile nature of salen-type ligands, this ligand class has been used to prepare model complexes for metalloenzymatic active sites and metallic cooperative effects.[1] Multinuclear complexes featuring salen-type ligands enable the study of magnetic and electronic interactions between the metal centers. Reinhoudt and co-workers synthesized several types of (ONNO)-crown-metallomacrocycles featuring various polyether chain lengths and diamine backbones (refer to chapter B.2.1).[11-13] These ligands stabilize a soft transition metal cation in the (ONNO) salen-type cavity and a hard alkali or alkaline earth metal cation in the polyether 18c6-type cavity with intermetallic distances of less than 4 Å. Co-complexation of a Lewis acidic metal center (Ba^{2+}, K^+, Na^+, Li^+) in the polyether cavity of a mononuclear transition metal complex produces a dinuclear complex and reduces the electron density at the phenolic oxygen atoms.[14-17] Concomitantly, the electron donating ability toward the transition metal is lowered, rendering the transition metal more positively charged and shifting its reduction potential to more positive values (anodic shift).

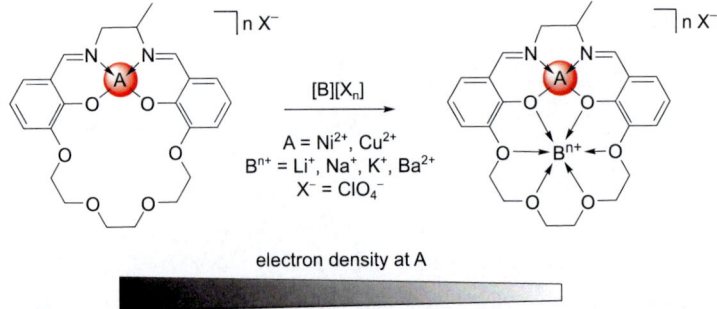

A = Ni^{2+}, Cu^{2+}
B^{n+} = Li^+, Na^+, K^+, Ba^{2+}
X^- = ClO_4^-

electron density at A

Scheme B.3.1. Decreased electronic density at the transition metal center A due to co-complexation of redox inactive Lewis acidic metal center B.[11, 14]

Results and Discussion

The electronic structures of metal complexes featuring salen-type ligands can be complex due to ambiguous electronic ground states.[1] These effects are often enhanced in multinuclear complexes featuring multiple salen-type subunits and multiple paramagnetic centers, which are magnetically coupled. Salen-type ligands are easily oxidized to produce ligand-centered radicals. The location of the oxidation event depends on the relative energies of the redox-active orbitals of the metal centers and the salen-type ligands. The oxidized complex may adopt any state within the continuum between an exclusively metal-centered oxidation affording a high-valent metal complex, and a ligand-centered oxidation producing a ligand radical complex. Due to the complexity of multimetallic salen systems, few investigations on the electronic structures have been reported. Glaser and co-workers determined the electrochemical and magnetic properties of several trinuclear complexes featuring triplesalen ligands.[1, 18] Upon oxidation of a trinickel(II) complex, valence tautomerism between a ligand centered radical and an oxidized metal center occurs (Scheme B.3.2).

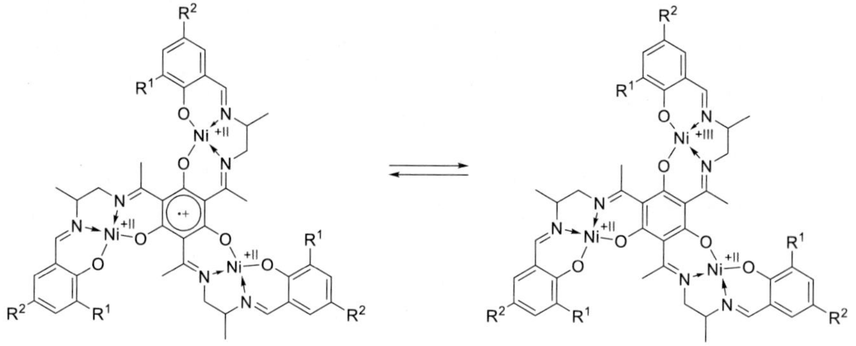

Scheme B.3.2. Valence tautomerism of a trinickel complex stabilized by a triplesalen ligand upon oxidation.[18]

Similar to triplesalen ligands, tris(ONNO) salen-type ligands may stabilize up to three redox active transition metal centers, giving access to multiple metal- or ligand-centered redox events. The properties of these complexes are modulated by an additional Lewis acidic metal cation in the central 18c6-type cavity. Tetranuclear heterometallic complexes featuring tris(ONNO)-type ligands may exhibit interesting electrochemical, electronic, magnetic and cooperative effects relevant to catalytic applications. Powell, Brooker and co-workers have thoroughly examined various lanthanide trizinc and tricopper complexes featuring tris(ONNO) salen-type macrocycles and determined their magnetic properties for potential application as single-molecule magnets.[19-24] Mashima, Okuda and co-workers have also utilized cooperative effects in lanthanide trizinc complexes for application in catalytic co-polymerization reactions.[25, 26] Depending on the central lanthanide metal cation, the catalytic activity may be enhanced.

B.3.2 Results and Discussion

Inspired by previous results on cooperative effects in multinuclear complexes featuring salen-type ligands, several heterometallic trizinc, trivanadyl and trimanganese complexes stabilized by salen tris(ONNO)-type ligands were synthesized.[19-26] Detailed spectroscopic, magnetic, and electrochemical studies were performed on these complexes and their structural properties were assessed.

B.3.2.1 Heterometallic Trizinc Complexes

For the preparation of multinuclear complexes featuring tris(ONNO)-type ligands, selective coordination of the metal centers at the designated binding sites is essential. Mononuclear complexes may be sufficiently stabilized against Schiff base rearrangements to enable subsequent metalation, affording heterometallic complexes. Heterometallic template synthesis or metalation of the salen-type binding sites of protonated pro-ligands have been used to prepare trizinc complexes.[19-21, 26-30] Here, the Zn^{2+} cations are selectively coordinated in the salen-type binding sites due to their small ionic radius of 0.6 Å (Shannon, coordination number 4).[31] Previous research has indicated that the tetranuclear calcium trizinc complex [$C^3LZn_3Ca(OAc)_2$] ([5$_{Ca}$][(OAc)$_2$]) with the C_3-bridging unit 2,2-dimethylpropane-1,3-diamine is accessible by heterometallic template synthesis.[25]

Since the trizinc system is well understood, it was used to establish the proof of principle that metalation of the mononuclear compounds selectively produces tetranuclear heterometallic complexes. First the targeted heterometallic calcium trizinc complex [$((R,R)$-$^{C2Ph2}L)Zn_3CaI_2$] ([11$_{Ca}$][I$_2$]) was prepared by the well-documented heterometallic template synthesis.[21-23, 26, 27, 30, 32] Pre-coordination of Zn^{2+} and Ca^{2+} cations to 3,6-diformylcatechol in boiling MeCN/MeOH (1:1) and subsequent condensation of the dialdehyde with (R,R)-1,2-diphenylethanediamine produced the [3+3] macrocycle (Scheme B.3.3). Alternatively, [11$_{Ca}$][I$_2$] was synthesized from [9$_{Ca}$][I$_2$] through deprotonation of the mononuclear starting complex with three equivalents of $Zn(OAc)_2 \cdot 2\ H_2O$. Both methods led to quantitative formation of the complex.

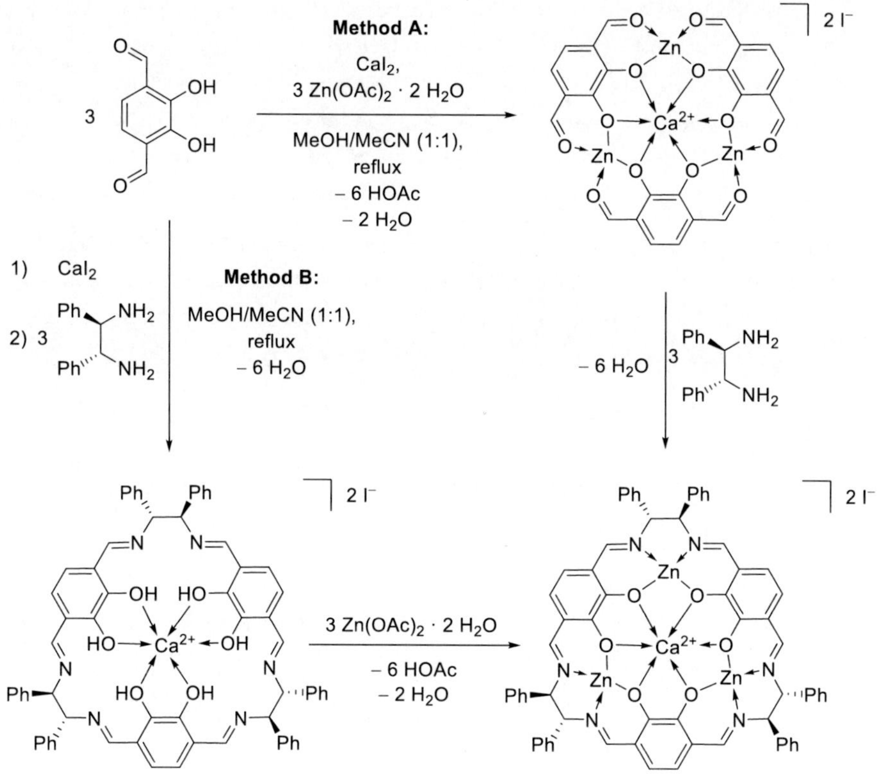

Scheme B.3.3. Synthesis of [11$_{Ca}$][I$_2$] either through heterometallic template synthesis (method A) or through deprotonation of the mononuclear complex [9$_{Ca}$][I$_2$] with Zn(OAc)$_2$ · 2 H$_2$O (method B).

The IR spectrum of [11$_{Ca}$][I$_2$] shows a prominent absorption of the imine bond at $\tilde{v}$ = 1622 cm^{-1}, which is shifted to lower wavenumbers compared to that of the mononuclear complex [9$_{Ca}$][I$_2$] at $\tilde{v}$ = 1630 cm^{-1}. The shift in the IR spectrum is similar to the one reported for [C^{3}LZn$_3$La(OAc)$_3$] ([5$_{La}$][(OAc)$_3$]) and indicates coordination of the salen-type binding sites to the zinc centers.[25, 26] The FAB mass spectrum shows a prominent peak at m/z = 1420.9, which can be assigned to the matrix adduct of the complex ([11$_{Ca}$][I$_2$] – HI – I$^-$ + 3 NBA), confirming the formation of the anticipated complex. Due to presumable π-π-stacking, the resolution of the FAB mass spectrum is very low. The complex is almost insoluble in most organic solvents and only slightly soluble in DMSO. Due to the low solubility, only ^{1}H NMR spectra were recorded. The ^{1}H NMR spectrum of [11$_{Ca}$][I$_2$] in DMSO-d_6 shows a singlet resonance for the imine protons at δ 8.01 ppm which is upfield shifted compared to the one for the mononuclear complex

[9$_{Ca}$][(OTf)$_2$] at δ 8.51 ppm. The signals of the catechol subunits are also upfield shifted to δ 6.20 ppm compared to δ 6.95 ppm for [9$_{Ca}$][(OTf)$_2$]. The resonances of the C$_2$-bridging unit at δ 5.27 ppm and for the phenyl protons at δ 7.42 – 7.25 ppm are broadened, presumably caused by a fluxional behavior of the C$_2$-bridging unit as has been suggested for the mononuclear complex [9$_{Ca}$][(OTf)$_2$] as well (section B.2.2). The ^{1}H NMR spectrum does not show any resonances that can be assigned to acetate anions, thus excluding scrambling of the anions.

Due to the low solubility of [11$_{Ca}$][I$_2$], the analogous complexes [((R,R)-^{C2Ph2}L)Zn$_3$Ca(OAc)$_2$] ([11$_{Ca}$][(OAc)$_2$]) and [((R,R)-^{C2Ph2}L)Zn$_3$CeCl$_3$] ([11$_{Ce}$][Cl$_3$]) were prepared in high yield using the heterometallic template synthesis with Zn(OAc)$_2$ · 2 H$_2$O and Ca(OAc)$_2$ or CeCl$_3$, respectively (Scheme B.3.4).

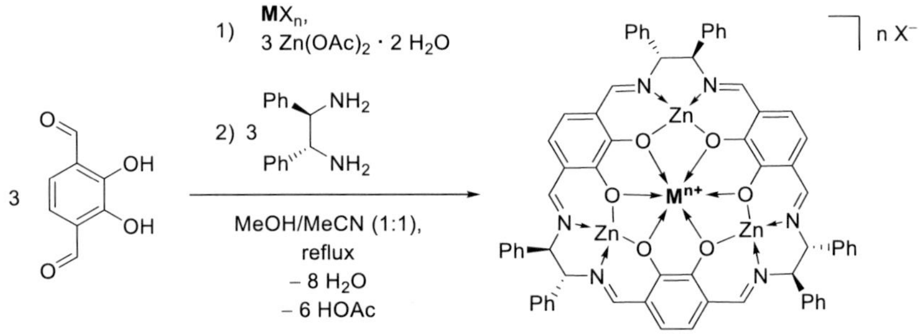

Scheme B.3.4. Heterometallic template synthesis of [11$_{Ca}$][(OAc)$_2$] and [11$_{Ce}$][Cl$_3$].

The FAB mass spectra of the complexes show prominent peaks at m/z = 1313.1 ([11$_{Ca}$][(OAc)$_2$]) and 1428.0 ([11$_{Ce}$][Cl$_3$]) which can be assigned to the [M – X$^-$]$^+$ fragments (X = OAc, Cl). The IR spectra of the complexes show absorptions of the imine bonds at $\tilde{\nu}$ = 1616 cm^{-1} ([11$_{Ca}$][(OAc)$_2$]) and 1619 cm^{-1} ([11$_{Ce}$][Cl$_3$]) which are shifted to lower wavenumbers similar to [11$_{Ca}$][I$_2$]. The complexes exhibit low solubility in most organic solvents and are only sparingly soluble in DMSO. The ^{1}H NMR spectrum of [11$_{Ce}$][Cl$_3$] in DMSO-d_6 shows two sets of signals for each expected resonance exhibited by the respective subunits of the macrocyclic ligand. The imine protons exhibit resonances at δ 8.76 and 8.46 ppm, and the protons of the catechol subunits at δ 6.03 and 5.64 ppm. The ^{1}H NMR resonances of the imine and catechol protons of [11$_{Ca}$][(OAc)$_2$] in DMSO-d_6 are upfield shifted similar to the ones of [11$_{Ca}$][I$_2$]. The

Results and Discussion

acetate anions exhibit a broad resonance at δ 1.78 ppm with a shoulder at δ 1.68 ppm which is upfield shifted compared to non-coordinating acetic acid at δ 1.91 ppm.[33] The C_2-bridging unit produces two broad resonances at δ 5.24 and 4.89 ppm which are presumably caused by fluxional behavior in solution.

A variable temperature ^{1}H NMR experiment was performed with [11$_{Ca}$][(OAc)$_2$] in DMSO-d_6 to further investigate the proposed exchange process (Figure B.3.3). Increasing the temperature from 297 K to 308 K, the two resonances of the acetate anions coalesce through an upfield shift of the resonance at δ 1.78 ppm and a downfield shift of the resonance at δ 1.68 ppm to produce a broad resonance at δ 1.72 ppm. VT NMR spectroscopy revealed that the signal width decreases from 88 Hz at 308 K to 9 Hz at 363 K, producing sharper resonances of the acetate anions. Similar behavior is observed for the signals of the imine protons at δ 7.92 ppm and for the catechol protons at δ 6.05 ppm. While increasing the temperature from 297 K to 323 K, the resonances of the protons of the C_2-bridging unit coalesce at δ 5.25 ppm through a downfield shift of the resonance at δ 4.92 ppm to δ 5.25 ppm, whereas the chemical shift of the signal at δ 5.25 ppm remains unchanged. Line shape analysis revealed that the broad resonance at δ 5.25 ppm becomes sharper at temperatures above 333 K and that the signal width decreases from 209 Hz at 323 K to 83 Hz at 363 K. Concomitantly, the signals exhibited by the protons of the phenyl substituents become sharper with increasing temperature and the expected coupling pattern becomes visible. The observed exchange process may be either caused by an exchange between coordinating and non-coordinating acetate or by conformational changes of the macrocyclic ligand backbone.

Heterometallic Tetranuclear Complexes Featuring a Tris(ONNO)-Type Ligand

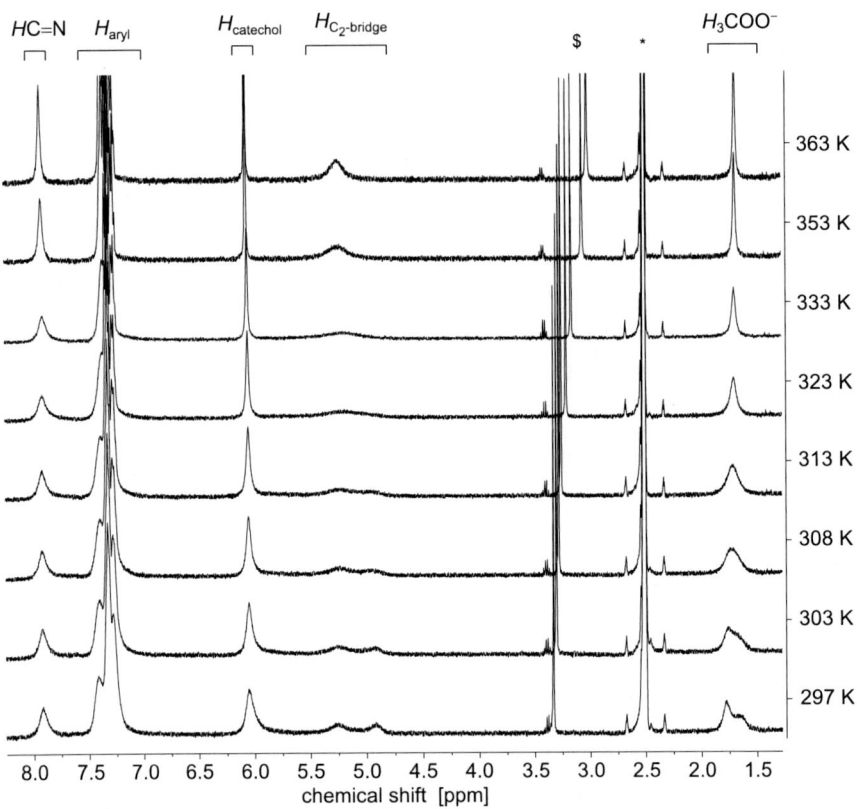

Figure B.3.3. VT ^{1}H NMR spectra of [11$_{Ca}$][(OAc)$_2$] in DMSO-d_6 in the temperature range of 297 – 363 K, referenced to residual (*) DMSO-d_5 with traces of water ($).

Single crystals of [11$_{Ca}$][(OAc)$_2$] were grown from a solution in DMSO-d_6 through slow evaporation of the solvent. The molecular structure was determined by X-ray diffraction analysis on single crystals (Figure B.3.4). Due to the low quality of the diffraction data set, only the zinc and calcium atoms were anisotropically refined. The unit cell contains two crystallographically independent molecules of [11$_{Ca}$(dmso)$_2$(H$_2$O)(OAc)$_2$] with nearly identical molecular geometry as well as additional 16 non-coordinating DMSO molecules. Similar to the mononuclear complex [9$_{Ca}$][(OTf)$_2$], the central calcium cation is coordinated in the 18c6-type cavity in distorted hexagonal bipyramidal fashion with the six catechol oxygen atoms located in the equatorial plane. In the apical positions, one molecule of DMSO and one molecule of water are coordinated to Ca^{2+}. The three zinc cations are exclusively located in the salen-type binding sites in distorted square-pyramidal fashion with τ_5 values of 0.12 (Zn1), 0.27 (Zn2) and 0.19 (Zn3). Zn1 and Zn2 are coordinated by acetate in η^1-coordination, whereas Zn3 features one apical DMSO molecule. Unlike the lanthanum trizinc complex ([5$_{La}$][(OAc)$_3$], the terminal acetate ligands are not bridged between zinc and calcium.[25, 26] The three zinc centers are located above the equatorial plane produced by the imine nitrogen and catecholate oxygen atoms (Figure B.3.4 (right)). The three square-pyramidal coordination polyhedra of the zinc centers, produced by the apical ligands, the equatorial oxygen and nitrogen atoms, are oriented in *syn*-configuration. The respective salen-type subunits adopt an umbrella conformation.[34]

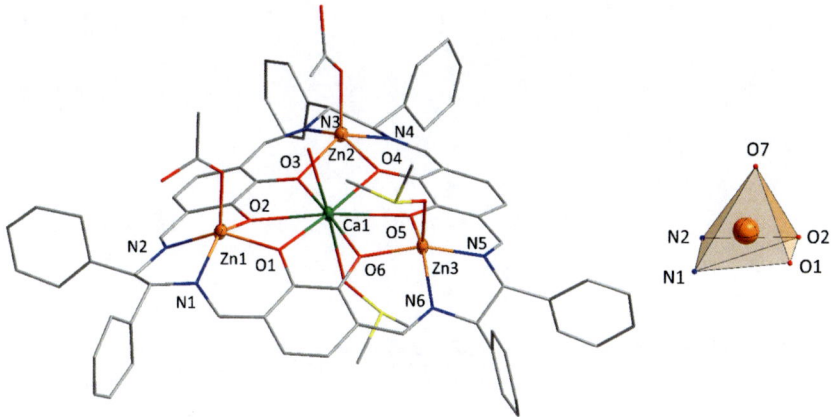

Figure B.3.4. (Left) Molecular structure of [11$_{Ca}$(dmso)$_2$(H$_2$O)(OAc)$_2$], 50% displacement ellipsoids for zinc and calcium; all hydrogen atoms and non-coordinating solvent molecules are omitted for clarity; (right) distorted square-pyramidal coordination polyhedron of Zn1.

Since trizinc complexes are selectively produced through deprotonation of the mononuclear complexes with three equivalents of Zn(OAc)$_2$, it was investigated whether binuclear heterometallic complexes can be prepared in a similar fashion. An equimolar solution of [9$_{Ca}$][I$_2$] and Zn(OAc)$_2$ · 2 H$_2$O in MeCN-d_3/MeOD-d_4 was heated for 30 min at 50 °C. ^{1}H NMR spectroscopy indicated the formation of multiple species (Figure B.3.5). Eight signals are observed in the region where the resonances of the imine protons would be expected. The resonances are tentatively assigned to unreacted starting complex (●) (*R*,*R*)-[H$_6$^{C2Ph2}LCaI$_2$], (■) monozinc complex [H$_4$((*R*,*R*)-^{C2Ph2}L)ZnCaI$_2$], (♦) dizinc complex [H$_2$((*R*,*R*)-^{C2Ph2}L)Zn$_2$CaI$_2$] and (○) trizinc complex [((*R*,*R*)-^{C2Ph2}L)Zn$_3$CaI$_2$]. The NMR data suggests that a binuclear complex is not selectively generated but instead complexes with varying nuclearity are produced.

Heterometallic Tetranuclear Complexes Featuring a Tris(ONNO)-Type Ligand

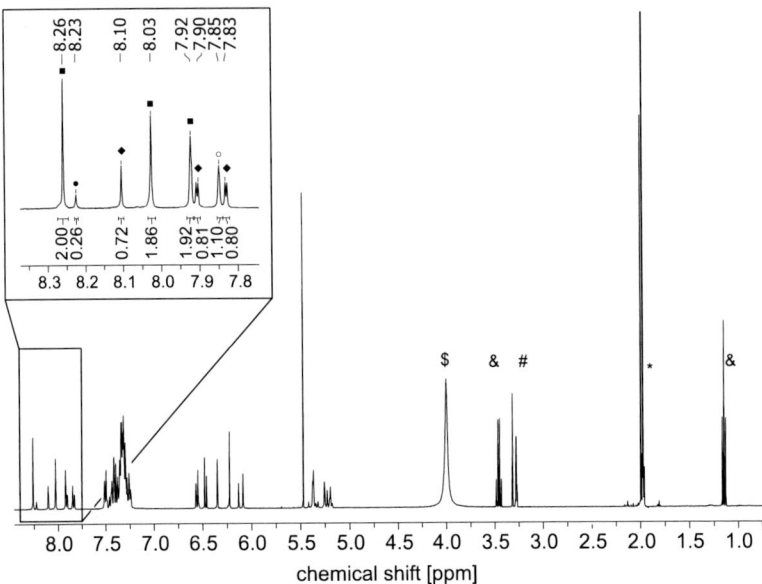

Figure B.3.5. ^{1}H NMR spectrum of the reaction of (R,R)-[H$_6^{C2Ph2}$LCaI$_2$] with Zn(OAc)$_2 \cdot$ 2 H$_2$O in MeCN-d_3/MeOD-d_4 referenced to residual solvent signals (*) MeCN-d_2 and MeOD-d_3 (#) with traces of (&) Et$_2$O and ($) H$_2$O; (●) starting complex, (■) [H$_4$((R,R)-^{C2Ph2}L)ZnCaI$_2$], (♦) [H$_2$((R,R)-^{C2Ph2}L)Zn$_2$CaI$_2$] and (○) [((R,R)-^{C2Ph2}L)Zn$_3$CaI$_2$].

Attempts to produce the analogous trimagnesium complexes [((R,R)-^{C2Ph2}L)Mg$_3$Ca(OAc)$_2$] in a similar approach to [11$_{Ca}$][(OAc)$_2$] were unsuccessful. The FAB mass spectrum of the reaction mixture showed a large number of peaks with low intensity, suggesting the formation of multiple species. The ^{1}H NMR spectrum of the reaction mixture in MeOD-d_4 only showed very broad resonances, thus confirming the formation of multiple species.

In this section the proof of concept was established that the central cation of the mononuclear complexes (section B.2) sufficiently stabilizes the [3+3] macrocycle to allow subsequent metalation of the salen-type binding sites affording heterometallic complexes. The well-established trizinc system was chosen to establish this proof of concept. The synthesis and characterization of the trizinc complexes [11$_{Ca}$][I$_2$], [11$_{Ca}$][(OAc)$_2$] and [11$_{Ce}$][Cl$_3$] was reported.

B.3.2.2 Heterometallic Trivanadyl Complexes

Since multiple trizinc complexes featuring tris(ONNO) salen-type ligands have been reported, the synthesis of tetranuclear complexes with redox active transition metal centers was investigated. Due to the rigidity of [3+3] macrocyclic ligands, three criteria were selected which may aid identifying potential candidates for selective complex formation.

Results and Discussion

1) To ensure that the transition metals are selectively coordinated in the salen-type binding sites, potential candidates should have a similar ionic radius to Zn^{2+} (0.6 Å).[31]
2) Molecular mononuclear or multinuclear complexes featuring salen-type ligands should be reported.
3) The salen-type binding sites need to be able to coordinate in square planar fashion to the metal centers.

Vanadium was selected as a potential candidate due to its rich redox chemistry which may be interesting for catalytic applications.[35-37] Pentacoordinate vanadium(IV) has a similar ionic radius of 0.53 Å to tetracoordinate Zn^{2+} and may be selectively coordinated at the salen-type binding sites.[31] Several vanadyl salen complexes have been reported in literature in which vanadium is coordinated in square pyramidal coordination geometry with the oxo ligand in the apical position.[38] In a first attempt, the direct heterometallic template synthesis similar to the one reported for ([5$_{La}$][(OAc)$_3$]) was tested. Performing the cyclization reaction between 3,6-diformylcatechol and 2,2-dimethylpropane-1,3-diamine with vanadyl acetylacetonate VO(acac)$_2$ and La(OAc)$_3 \cdot$ H$_2$O as templating cations, selective complex formation was not observed. The FAB mass spectrum of the reaction mixture showed multiple signals with a low signal to noise ratio, indicating the formation of multiple undefined species.

Since the proof of concept was established that heterometallic complexes can be prepared from mononuclear ones, this approach was used to prepare trivanadyl complexes (section B.3.2.1). Ligand exchange reaction between three equivalents of VO(acac)$_2$ and [7$_{Ca}$][(OTf)$_2$] in MeOH under inert gas conditions afforded the tetranuclear calcium trivanadyl complex [((R,R)-^{Cy}L)(VO)$_3$Ca(OTf)$_2$] ([12$_{Ca}$][(OTf)$_2$]) in high yield (Scheme B.3.5). The heterometallic complex was obtained as an orange/brown solid that is moderately air sensitive: The compound decomposes over several days under air to produce dark insoluble residues.

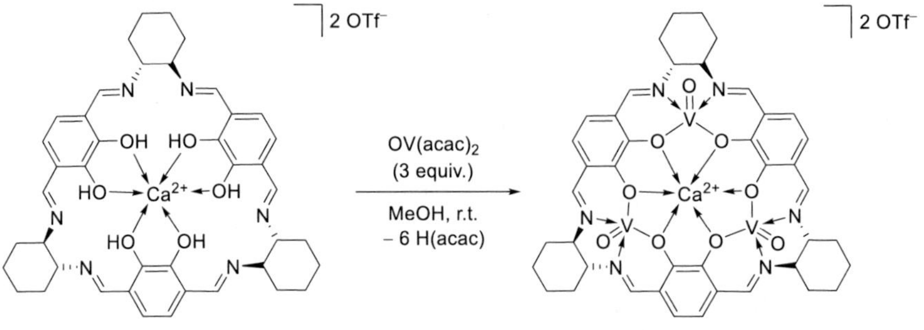

Heterometallic Tetranuclear Complexes Featuring a Tris(ONNO)-Type Ligand

Scheme B.3.5. Ligand exchange reaction between three equivalents of VO(acac)$_2$ and [7$_{Ca}$][(OTf)$_2$] to produce the calcium trivanadyl complex [12$_{Ca}$][(OTf)$_2$].

[12$_{Ca}$][(OTf)$_2$] is paramagnetic and does not show any resonances in its ^{1}H NMR spectrum. The absorption of the imine bond in the IR spectrum is shifted from $\tilde{v}$ = 1637 cm^{-1} ([7$_{Ca}$][(OTf)$_2$]) to 1630 cm^{-1} ([12$_{Ca}$][(OTf)$_2$]), indicating successful coordination of the salen-type binding sites to vanadium (Figure B.3.6). The IR spectrum also shows a new vibrational mode at $\tilde{v}$ = 996 cm^{-1} which is in a typical region for V=O vibrational modes.[39] The FAB mass spectrum shows a major peak at 1116.0 which can be assigned to the [M – OTf]$^+$ fragment confirming formation of the calcium trivanadyl complex.

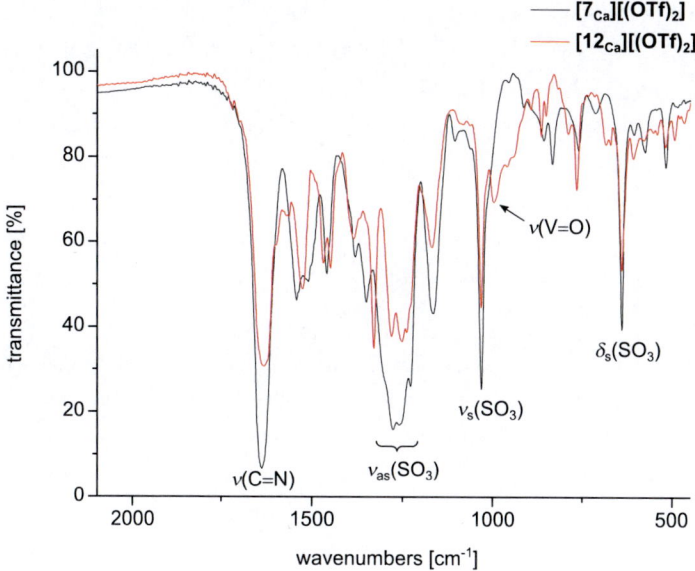

Figure B.3.6. Excerpt of the IR spectra of [7$_{Ca}$][(OTf)$_2$] and [12$_{Ca}$][(OTf)$_2$] measured on a KBr pellet.

[12$_{Ca}$][(OTf)$_2$] is soluble in MeCN/MeOH (1:1) and DMSO but insoluble in THF, Et$_2$O and hydrocarbons. Due to the low solubility of [12$_{Ca}$][(OTf)$_2$], the analogous complex featuring the 1,2-diphenylethylenediamine bridging unit was prepared. [((R,R)-^{C2Ph2}L)(VO)$_3$Ca(OTf)$_2$] ([13$_{Ca}$][(OTf)$_2$]) was prepared similar to [12$_{Ca}$][(OTf)$_2$] through ligand exchange reaction between three equivalents of VO(acac)$_2$ and [9$_{Ca}$][(OTf)$_2$] in MeCN/MeOH (1:1) under inert gas conditions (Scheme B.3.6). Following purification, the complex was obtained as an orange, air sensitive solid in high yield.

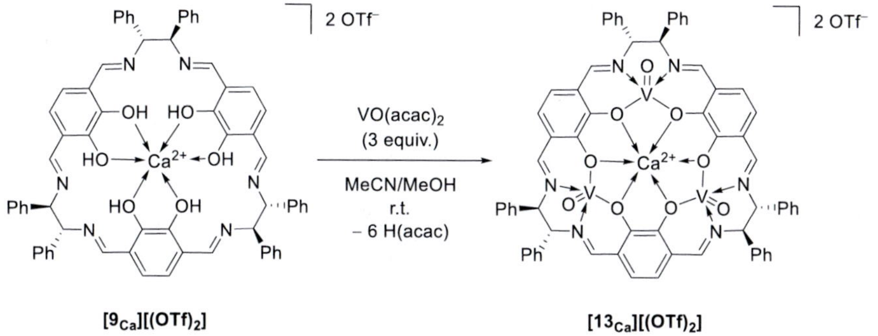

Scheme B.3.6. Ligand exchange reaction between three equivalents of VO(acac)$_2$ and [9$_{Ca}$][(OTf)$_2$] to produce the calcium trivanadyl complex [13$_{Ca}$][(OTf)$_2$].

[13$_{Ca}$][(OTf)$_2$] is soluble in DCM, acetonitrile, DMF and MeOH but insoluble in THF, Et$_2$O and hydrocarbons. The IR spectrum of [13$_{Ca}$][(OTf)$_2$] shows a shift of the imine vibrational mode from $\tilde{v}$ = 1632 cm^{-1} ([9$_{Ca}$][(OTf)$_2$]) to 1616 cm^{-1} ([13$_{Ca}$][(OTf)$_2$]), indicating coordination of the salen-type binding sites to the vanadium centers. The vibrational mode of the V=O bond is observed at $\tilde{v}$ = 1000 cm^{-1}, which is in the typical range for terminal V=O bonds with square-pyramidal coordination geometry of the vanadium center.[39] The FAB mass spectrum shows a peak at m/z = 1260.0 which can be assigned to [M − 2OTf$^-$]$^+$, indicating formation of the calcium trivanadyl complex. Similar to [12$_{Ca}$][(OTf)$_2$], the complex is paramagnetic and does not produce any resonances in the ^{1}H NMR spectrum.

EPR spectroscopy was used to gain further insight into the chemical environment of the vanadium centers. Coupling of the electron spin with the ^{51}V nucleus (I = 7/2) produces an octet signal (2I + 1 = 8 lines) in the X-band EPR spectrum of [13$_{Ca}$][(OTf)$_2$] in MeCN at 298 K (Figure B.3.7(a)).[40] Due to the strong V=O interaction, an axial dependency of the molecule to the external field is observed, producing a slightly anisotropic EPR signal.[41] In case of an isotropic vanadium center, the intensity of all lines would be the same. The hyperfine coupling constant of A = 292 MHz and the effective g-factor of g_{eff} = 1.96 were determined through fitting of the experimental EPR spectrum with an isotropic vanadium d^1 model (Figure B.3.7 (b)). These values are in a typical range for other mononuclear vanadyl salen complexes.[38, 41] Maeda and co-workers reported that the mononuclear [(salen)VO] (salen = bis(salicylidene)ethylenediamine) complex in DMSO produces an anisotropic EPR spectrum at 77 K with g-values of $g_\parallel$ = 1.957 and $g_\perp$ = 1.988, and hyperfine coupling constants of $A_\parallel$ = 491 MHz and $A_\perp$ = 188 MHz.[38] These values correspond to the isotropic g_{iso}-value with the expression g_{iso} = ($g_\parallel$ + 2 $g_\perp$)/3 = 1.98 and the isotropic hyperfine coupling constant with the expression A_{iso} = ($A_\parallel$ + 2 $A_\perp$)/3 = 289 MHz. These isotropic values are similar to the ones obtained for [13$_{Ca}$][(OTf)$_2$] and, thus, substantiate the assignment of the EPR signal of [13$_{Ca}$][(OTf)$_2$] to three equivalent vanadium d^1 centers. However, the EPR spectrum of [13$_{Ca}$][(OTf)$_2$] also shows a smaller second signal that indicates a certain degree of inequivalence of the vanadium centers or formation of a second species.

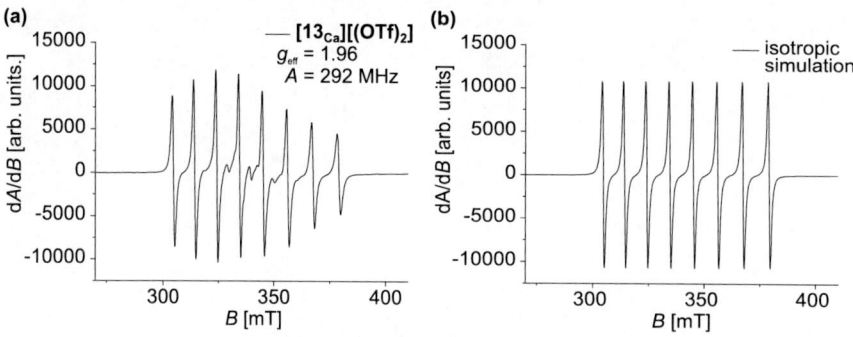

Figure B.3.7. (a) Experimental X-band EPR spectrum of [13$_{Ca}$][(OTf)$_2$] in MeCN at 298 K and (b) isotropic simulation of the EPR spectrum.

EPR and IR spectroscopy gave insight into the coordination environments of the respective vanadium centers, indicating retention of the +IV oxidation state and coordination of vanadium in the salen-type binding sites. However, these spectroscopic methods fail to give insight into the overall properties of the tetranuclear complexes. Two different configurations may be produced in which the oxo ligands are either in *syn-* (up-up-up) or *anti-*configuration (up-down-up). Additionally, the three vanadium centers are coordinated in close proximity to each other that may produce metallic cooperative effects due to magnetic coupling of the metal spins.

To further understand the overall electronic properties of the complex, the magnetic data of [13$_{Ca}$][(OTf)$_2$] were recorded with a SQUID magnetometer. The magnetic data of [13$_{Ca}$][(OTf)$_2$] are presented as $\chi_m T$ as function of the temperature at 0.1 T and as molar magnetization (M_m) vs. the magnetic field B (Figure B.3.8). The $\chi_m T$ value of 1.09 cm^3 K mol^{-1} (μ_{eff} = 2.91 μ_B) at 290 K is at the lower limit of the range 1.06 – 1.19 cm^3 K mol^{-1} for three non-interacting vanadium(IV) centers with an overall spin of S_{total} = 3/2. The magnetic data at 290 K thus confirms the coordination of three vanadium d^1 centers per complex. The $\chi_m T$ values remain approximately constant in the temperature range 100 – 290 K. At temperatures below 100 K, the values drop to 0.384 cm^3 K mol^{-1} at 2.0 K. Since vanadyl(IV) (VO^{2+}) units may be described as approximately isotropic spin centers with an effective spin of S = 1/2. Simulation of the data suggests that the drop-off is almost exclusively caused by predominant antiferromagnetic exchange interactions (*vide infra*). To a minor degree the drop-off is due to single-ion effects causing small anisotropy of the vanadium centers. The molar magnetization shows an almost linear dependence on the magnetic field up to 5.0 T at 2.0 K. At 5.0 T, the molar magnetization M_m is 1.3 N_A μ_B and not saturated. Additionally, the graph shows a small change of the curvature hinting at an inflexion point of the M_m vs. B curve at 3.0 to 3.5 T. Since the inflexion point occurs at 0.9 – 1.0 N_A μ_B and since saturation steps may occur at g_{eff} · 1/2 N_A μ_B and g_{eff}

· 3/2 $N_A \mu_B$ ($g_{eff} \approx 2$) for a system of three $S = 1/2$ centers, the inflexion potentially indicates a change of the ground state of the three-center-system from $S_{total} = 1/2$ to $3/2$ at 2.0 K.

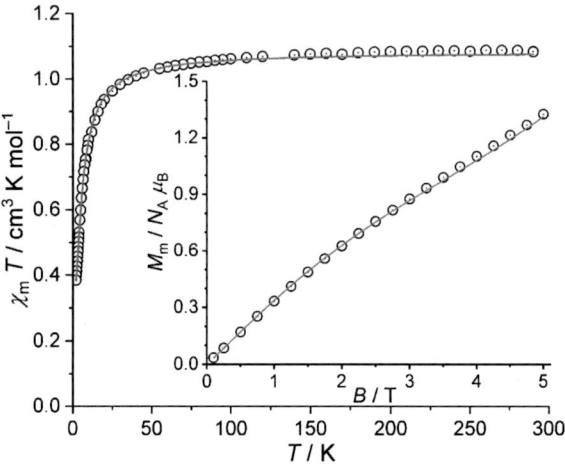

Figure B.3.8. Magnetic data of **[13Ca][(OTf)₂]** with temperature dependence of $\chi_m T$ at 0.1 T; insert: molar magnetization M_m vs. applied field B at 2.0 K; experimental data (open circles), least-square fit using an effective spin model with three interacting $S = 1/2$ centers.

To quantify the exchange interactions, an effective isotropic spin model using the effective spin option of CONDON was used.[42, 43] The corresponding Hamiltonian with the exchange interaction parameters J_{ij}, the effective g-factor g_{eff}, and the spin operators $\hat{S}_i$ representing the three vanadium(IV) centers is depicted in (Equation B.3.1).

$$\hat{H} = -2J_{12}\hat{S}_1 \cdot \hat{S}_2 - 2J_{23}\hat{S}_2 \cdot \hat{S}_3 - 2J_{13}\hat{S}_1 \cdot \hat{S}_3 + g_{eff}\mu_B \sum_{i=1}^{3} B \cdot \hat{S}_i \qquad \text{(Equation B.3.1)}$$

Least-squares fitting of the data with a relative mean squared error of 1% yields $J_{12} = J_{23} = J_{13} = (-1.84 \pm 0.09)$ cm⁻¹ and $g_{eff} = 1.96 \pm 0.02$. The effective g-factors obtained by SQUID measurement and EPR spectroscopy (Figure B.3.7) are identical within the margin of errors and indicate the consistency of the spectroscopic and magnetic data. The small deviation of the fit from the data is due to the applied corrections and the small anisotropy of the vanadium(IV) centers, which was neglected in the fitting process. The small anisotropy is also observed in the corresponding EPR spectrum (Figure B.3.7). The exchange interaction parameters indicate weak antiferromagnetic exchange interactions between the three vanadium(IV) centers. Due to the rather large distances between the three vanadium centers in the macrocyclic complex, only a weak interaction is expected. The ground state of the three centers is characterized by $S_{total} = 1/2$ at 2.0 K. The effective g-factor is representative of similar centers.[44] The antiferromagnetic exchange interactions and the almost linear shape of the magnetization curve with very weak features indicate that this system is another example of a frustrated spin system. Due to the geometrical constrains of the macrocyclic ligand, the three vanadium centers, all with $S = 1/2$, produce a triangular shape (Figure B.3.9). If all spins had antiparallel orientation to the neighboring spins, the energy of the spin system would be minimum. If two spins are aligned antiparallel to each other, the third spin is frustrated, since both spin orientations will result in the same energy. The third spin is not able to simultaneously minimize the interaction with the other two spins. Since spin frustration occurs for each center, the ground state is sixfold degenerate. The frustrated spin system rationalizes the antiferromagnetic exchange interactions.

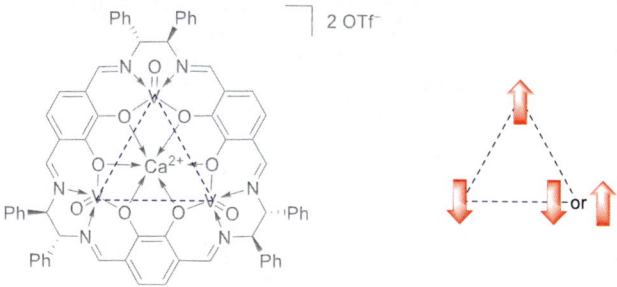

Figure B.3.9. Schematic illustration of the frustrated spin system in [13$_{Ca}$][(OTf)$_2$].

To probe whether oxidation of [13$_{Ca}$][(OTf)$_2$] may lead to selective formation of a diamagnetic complex in which all vanadium(IV) centers are oxidized to vanadium(V), the electrochemical properties of the complex were determined by cyclic voltammetry (Figure B.3.10). The cyclic voltammogram of [13$_{Ca}$][(OTf)$_2$] in acetonitrile reveals several broad, mostly irreversible oxidation events in the range of $E = 0.25 - 1.50$ V. Two very broad irreversible reductions occur at $E = -1.00$ V and at $E = -1.74$ V. Due to the known redox non-innocence of salen-type ligands, the observed redox events cannot be undoubtedly assigned to metal- or ligand-centered processes.[1, 45] In addition to the three reduction events of the vanadium centers, up to nine ligand-centered oxidation events may potentially occur: Oxidation of the three catecholate subunits to quinones via semi-quinones (three two-electron-oxidation reactions) and oxidation of the three vanadium(IV) centers to vanadium(V) (three one-electron-oxidation reactions). Due to the large number of potential redox events, further electrochemical characterizations were not performed.

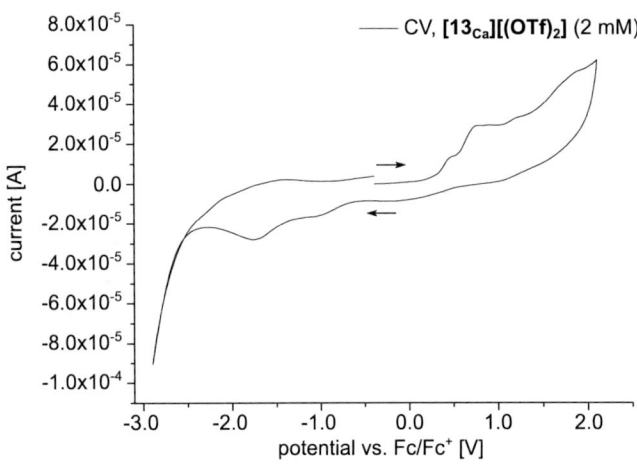

Figure B.3.10. Cyclic voltammogram of [13$_{Ca}$][(OTf)$_2$] (2 mM) in MeCN at 298 K, electrolyte [^{n}Bu$_4$N][PF$_6$] (100 mM), referenced to the Fc/Fc$^+$ couple.

To further elucidate the influence of the anion on the properties of the trivanadyl complexes and to expand the pool of potential catalysts, calcium trivanadyl complexes [((R,R)-^{C2Ph2}L)(VO)$_3$CaX$_2$] with X = Cl for [13$_{Ca}$][Cl$_2$], Br for [13$_{Ca}$][Br$_2$] or I for [13$_{Ca}$][I$_2$] were prepared similar to [13$_{Ca}$][(OTf)$_2$] from the respective mononuclear complexes (Scheme B.3.7). The heterometallic calcium trivanadyl complexes were obtained in high yield after purification. The complexes are paramagnetic and do not produce any resonances in the ^{1}H NMR spectra. The IR spectra of the complexes confirm the formation of the trivanadyl complexes as indicated by absorptions in the range of $\tilde{v}$ = 1614 – 1616 cm^{-1} for the imine bonds and at $\tilde{v}$ = 999 – 1002 cm^{-1} for the V=O bonds. The FAB mass spectra show peaks at m/z = 1296.0 ([13$_{Ca}$][Cl$_2$], [M – X$^-$]$^+$), 1425,8 ([13$_{Ca}$][Br$_2$], [M]$^+$) and 1386.9 ([13$_{Ca}$][I$_2$], [M – X$^-$]$^+$), indicating successful formation of the complexes.

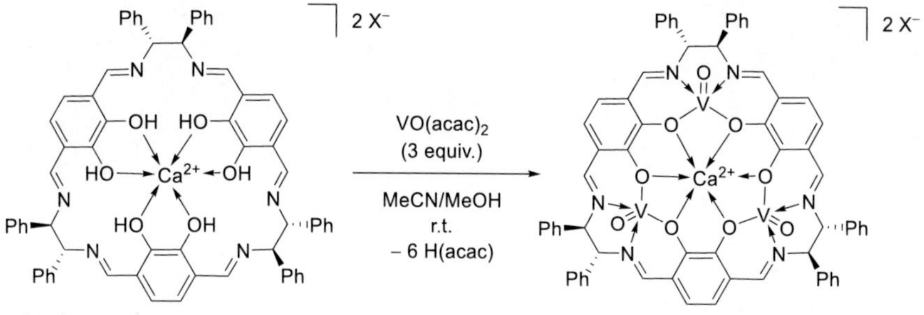

[9_{Ca}][X_2]

X = Cl: **[13_{Ca}][Cl_2]**

X = Br: **[13_{Ca}][Br_2]**

X = I: **[13_{Ca}][I_2]**

Scheme B.3.7. Synthesis of the calcium trivanadyl complexes **[13_{Ca}][Cl_2]**, **[13_{Ca}][Br_2]** and **[13_{Ca}][I_2]**.

EPR spectra of the compounds were recorded at 298 K in acetonitrile (Figure B.3.11). The EPR spectrum of **[13_{Ca}][Cl_2]** (Figure B.3.11(a)) shows an octet signal with a small degree of anisotropy similar to the EPR signal of **[13_{Ca}][$(OTf)_2$]** (Figure B.3.11(b)). The effective *g*-factor g_{eff} = 1.96 and the hyperfine coupling constant of A = 292 MHz are identical to the parameters for **[13_{Ca}][$(OTf)_2$]**. However, the contribution of the minor signal to the overall EPR signal of **[13_{Ca}][Cl_2]** is increased compared to **[13_{Ca}][$(OTf)_2$]**. In the case of the bromide and iodide complexes, the contribution of the second signal to the EPR signal becomes predominant. Due to superposition of the signals, the effective *g*-factors and hyperfine coupling constants cannot be determined (Figure B.3.11(c), (d)). Since the contribution of the second signal to the EPR spectrum becomes more predominant when exchanging the halogen anion with the higher homologues, the data suggests an increasing interaction of the anion with vanadium when going from chloride to iodide, producing inequivalent vanadium centers.

Heterometallic Tetranuclear Complexes Featuring a Tris(ONNO)-Type Ligand

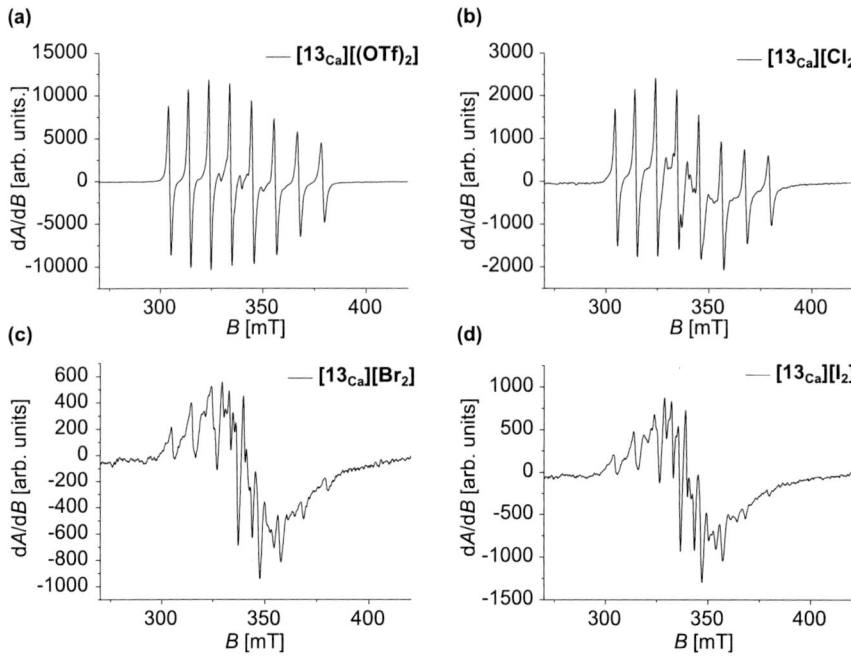

Figure B.3.11. X-band EPR spectra of (a) [13$_{Ca}$][(OTf)$_2$], (b) [13$_{Ca}$][Cl$_2$], (c) [13$_{Ca}$][Br$_2$] and (d) [13$_{Ca}$][I$_2$] in MeCN at 298 K.

To determine the influence of the central metal cation in the 18c6-type cavity onto the properties of the trivanadyl complexes, [((R,R)-^{C2Ph2}L)(VO)$_3$Sr(OTf)$_2$] ([13$_{Sr}$][(OTf)$_2$]), [((R,R)-^{C2Ph2}L)(VO)$_3$Ba(OTs)$_2$] ([13$_{Ba}$][(OTs)$_2$]), [((R,R)-^{C2Ph2}L)(VO)$_3$La(OTf)$_3$] ([13$_{La}$][(OTf)$_3$]) and [((R,R)-^{C2Ph2}L)(VO)$_3$CeCl$_3$] ([13$_{Ce}$][Cl$_3$]) were prepared similar to [13$_{Ca}$][(OTf)$_2$] from the mononuclear complexes (Scheme B.3.8). The tetranuclear trivanadyl complexes were obtained in high yield after purification. The complexes are paramagnetic and do not produce any resonances in their respective ^{1}H NMR spectra. The IR spectra of the complexes confirm formation of the trivanadyl complexes as indicated by absorptions of the imine bonds at $\tilde{v}$ = 1614 – 1622 cm^{-1} and of the V=O bonds at $\tilde{v}$ = 1001 – 1003 cm^{-1}. The FAB mass spectra show peaks at m/z = 1456.6 ([13$_{Sr}$][(OTf)$_2$]), 1529.4 ([13$_{Ba}$][(OTs)$_2$]) and 1656.3 ([13$_{La}$][(OTf)$_3$]) corresponding to the [M – X$^-$]$^+$ fragments (X = OTf, OTs), and m/z = 1461.2 ([13$_{Ce}$][Cl$_3$]) corresponding to the [M – e$^-$]$^+$ fragment.

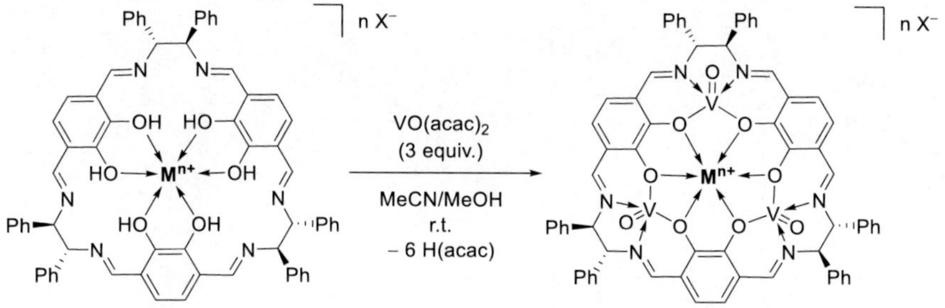

Results and Discussion

Scheme B.3.8. Synthesis of the tetranuclear trivanadyl complexes [13$_{Sr}$][(OTf)$_2$], [13$_{Ba}$][(OTs)$_2$], [13$_{La}$][(OTf)$_3$], [13$_{Ce}$][Cl$_3$].

EPR spectra of the compounds were recorded at 298 K in acetonitrile (Figure B.3.12). Similar to [13$_{Ca}$][(OTf)$_2$], the complexes [13$_{Ba}$][(OTs)$_2$] and [13$_{La}$][(OTf)$_3$] produce a slightly anisotropic octet signal at an effective g-factor of g_{eff} = 1.96 with a hyperfine coupling constant of A = 292 MHz. In the case of the analogous strontium complex [13$_{Sr}$][(OTf)$_2$], the EPR spectrum is similar to theose of [13$_{Ca}$][Br$_2$] and [13$_{Ca}$][I$_2$]: The contribution of the second signal to the EPR spectrum becomes predominant and the effective g-factors and hyperfine coupling constants cannot be determined (Figure B.3.12(b)). The second signal may be produced by inequivalent vanadium centers due to a different ligand conformation. For the cerium complex [13$_{Ce}$][Cl$_3$], only a weak EPR signal with low intensity is detected due to superposition of the vanadyl and cerium signals.

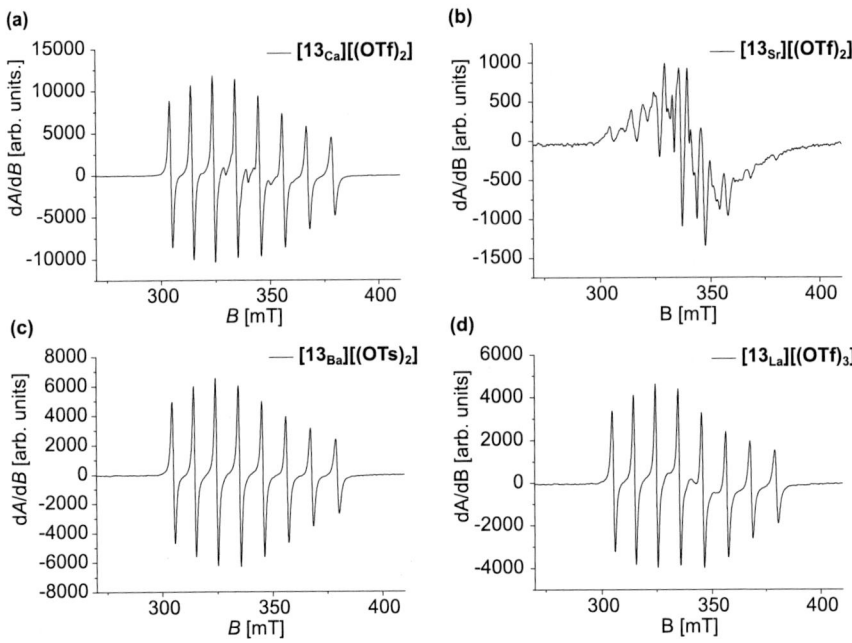

Figure B.3.12. X-band EPR spectra of **[13$_{Ca}$][(OTf)$_2$]**, **[13$_{Sr}$][(OTf)$_2$]**, **[13$_{Ba}$][(OTs)$_2$]** and **[13$_{La}$][(OTf)$_3$]** in acetonitrile at 298 K.

Attempts to produce the structurally similar but diamagnetic complex [((R,R)-C2Ph2L)(TiO)$_3$Ca(OTf)$_2$] (**[14$_{Ca}$][(OTf)$_2$]**) in a similar approach to the calcium trivanadyl analogue were inconclusive. **[14$_{Ca}$][(OTf)$_2$]** was prepared similar to **[13$_{Ca}$][(OTf)$_2$]** from **[9$_{Ca}$][(OTf)$_2$]** and TiO(acac)$_2$. The FAB mass spectrum shows a peak at m/z = 1401.1 which can be assigned to the fragment [M − OTf⁻]. IR spectroscopy revealed the characteristic absorption of the imine bond at $\tilde{\nu}$ = 1638 cm^{-1} which is shifted to higher wavenumbers, unlike the analogous trivanadyl complex **[13$_{Ca}$][(OTf)$_2$]** at $\tilde{\nu}$ = 1616 cm^{-1}. The ^{1}H NMR spectrum of **[14$_{Ca}$][(OTf)$_2$]** in DMSO-d_6 showed only very broad resonances. The spectroscopic data suggests formation of multiple species, whereas mass spectrometry indicated formation of the anticipated complex **[14$_{Ca}$][(OTf)$_2$]**. Most likely, **[14$_{Ca}$][(OTf)$_2$]** is produced in low selectivity together with multiple other species, producing broad resonances in the ^{1}H NMR spectrum.

In this section, the approach from section B.3.2.1 was successfully used to prepare mixed heterometallic complexes featuring macrocyclic ligands. Several tetranuclear trivanadyl complexes featuring alkaline earth metal or lanthanide metal cations in the 18c6-type cavity were synthesized and characterized.

Results and Discussion

B.3.2.3 Heterometallic Trimanganese Complexes

Structural comparison of calcium trimanganese complexes stabilized by a macrocyclic tris(ONNO) ligand with the [Mn_3CaO_4] cubane model by Agapie and co-workers may give valuable insights into the properties of the natural occurring [Mn_4CaO_n] cluster.[7-10] Due to the rich redox chemistry of manganese, heterometallic manganese complexes may also be potential candidates for catalytic redox reactions. Tetracoordinate manganese(II) has a similar ionic radius of 0.66 Å to tetracoordinate Zn^{2+} (0.6 Å) and may thus be coordinated at the salen-type binding sites.[31] Most manganese salen complexes feature the metal in the formal oxidation state +III and are applied as Jacobson's catalyst in epoxidation reactions.[46] Previous studies focused on direct heterometallic template synthesis to produce lanthanum trimanganese complexes which were not selectively synthesized by this approach.[25]

In this work, the synthesis of trimanganese complexes starting from mononuclear macrocyclic complexes was investigated. In a first approach, [6_{Ca}][(OTf)$_2$] and [9_{Ca}][(OTf)$_2$] were treated with three equivalents of Mn(OAc)$_2$ · 4 H$_2$O in MeOH or MeCN/MeOH (1:1) under air, similar to the preparation of Jacobson's catalyst to produce a manganese(III) complex. The FAB mass spectrum of the reaction mixture showed a low signal to noise ratio with multiple signals, indicating formation of multiple undefined species. Most likely manganese is oxidized during the reaction, producing organic radicals which partially oxidize the salen-type subunits and decompose the ligand scaffold.

In a second approach, the synthesis of a trimanganese(II) complex was investigated. To avoid potential side reactions due to formation of acetic acid and water when using Mn(OAc)$_2$ · 4 H$_2$O, anhydrous MnCl$_2$ was selected. *In situ* deprotonation of [9_{Ca}][(OTf)$_2$] in MeCN/THF (1:1) with six equivalents of LiHMDS to produce the intermediate complex [((R,R)-^{C2Ph2}L)Li$_6$Ca(OTf)$_2$] ([15_{Ca}][(OTf)$_2$]) and subsequent salt metathesis reaction with anhydrous MnCl$_2$ resulted in selective formation of the complex [((R,R)-^{C2Ph2}L)(Mn(II)Cl)$_2$(Mn(III)Cl)Ca(thf)$_2$] ([16_{Ca}][Cl$_3$]). Presumably, the third Mn(II) center is oxidized to Mn(III) during workup with DCM but the precise mechanism remains unclear. The complex was obtained as a moisture- and air-sensitive orange solid in high yield after purification. [16_{Ca}][Cl$_3$] decomposes within a few seconds under air to produce a dark black residue. The complex is soluble in THF, DCM, chloroform and MeCN, but insoluble in hydrocarbons and Et$_2$O. A satisfactory elemental composition could not be determined by combustion analysis, presumably due to carbide formation. Atomic absorption spectroscopy revealed a lower calcium and a higher manganese content than expected, presumably due to matrix effects of the analyte. The manganese and calcium content correspond to one calcium and three manganese atoms per molecule. Determination of the chloride content by ion chromatography confirms three chlorides per molecule. Isolation of the *in situ* deprotonated complex [15_{Ca}][(OTf)$_2$] was unsuccessful.

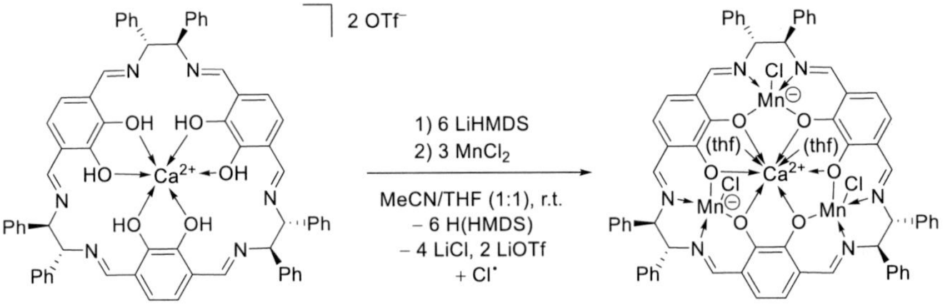

Heterometallic Tetranuclear Complexes Featuring a Tris(ONNO)-Type Ligand

Scheme B.3.9. In situ deprotonation of [9$_{Ca}$][(OTf)$_2$] with six equivalents LiHMDS to produce the intermediate [15$_{Ca}$][(OTf)$_2$], and subsequent salt metathesis with three equivalents of MnCl$_2$ to produce [16$_{Ca}$][Cl$_3$]. The precise mechanism of the one-electron oxidation of one manganese(II) center to a manganese(III) center remains unclear.

[16$_{Ca}$][Cl$_3$] is paramagnetic and does not produce any resonances in the ^{1}H NMR spectrum. The IR spectrum of [16$_{Ca}$][Cl$_3$] shows a shift of the absorption of the imine bond from $\tilde{\nu}$ = 1632 cm^{-1} ([9$_{Ca}$][(OTf)$_2$]) to 1616 cm^{-1} ([16$_{Ca}$][Cl$_3$]). The shift is similar to the one observed for the trivanadyl complex [13$_{Ca}$][(OTf)$_2$] and indicates coordination of the salen-type binding sites to the transition metal. Since the characteristic absorptions of the triflate anion at $\tilde{\nu}$ = 1300 – 1150 cm^{-1} (ν_{as}(SO$_3$)), 1031 cm^{-1} (ν_s(SO$_3$)) and 638 cm^{-1} (δ_s(SO$_3$)) disappeared, the IR spectrum suggests an anion exchange reaction. The FAB mass spectrum of [16$_{Ca}$][Cl$_3$] shows major peaks at m/z = 1260.1 and 1295.0 which can be assigned to the chloride containing fragments [M – 2 Cl$^-$] and [M – Cl$^-$], respectively. The ESI mass spectrum shows a major peak at m/z = 1315.06958 which can be assigned to [M – Cl$^-$ + H$_3$O]$^+$. Mass spectrometry confirms exchange of the triflate anions against chloride as was suggested by IR spectroscopy.

Single crystals of [16$_{Ca}$][Cl$_3$] were grown from THF/MeCN through vapor diffusion of Et$_2$O into the solution. The molecular structure was determined by single crystal X-ray diffraction analysis (Figure B.3.13). Due to disordered solvent molecules in the lattice of [16$_{Ca}$][Cl$_3$], the diffraction data was treated with the SQUEEZE routine.[47] Similar to the mononuclear complex [9$_{Ca}$][(OTf)$_2$], the central calcium atom is coordinated in the 18c6-type cavity in distorted hexagonal bipyramidal fashion: The six catechol oxygen atoms are located in the equatorial plane and two molecules of THF are coordinated in the apical positions. The three manganese centers are exclusively located in the salen-type binding sites and coordinated in distorted square pyramidal coordination geometry. The chloro ligands are located in the apical positions and adopt *syn*-configuration. The salen-type subunits adopt an umbrella conformation, similar to the trizinc analogue (Figure B.3.4). The coordination polyhedra of the three manganese centers are depicted in Figure B.3.13(b) – (d).

Results and Discussion

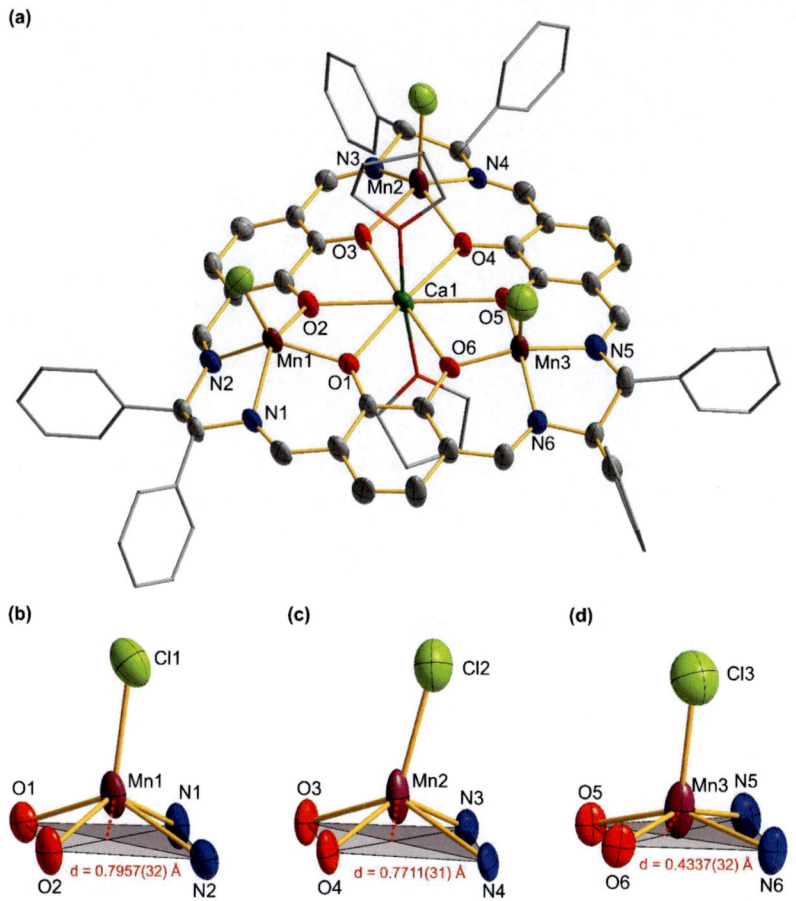

Figure B.3.13. (a) Molecular structure of **[16$_{Ca}$][Cl$_3$]** with 50% displacement ellipsoids, all hydrogen atoms and non-coordinating solvent molecules are omitted for clarity. Selected bond lengths (Å): Mn1–Cl1 2.359(2), Mn2–Cl2 2.354(2), Mn3–Cl3 2.352(3). Coordination geometries of (b) Mn1, (c) Mn2 and (d) Mn3, only atoms coordinated to manganese are shown. Calculated distances (Å) of the respective manganese centers to the plane produced by the coordinating oxygen and nitrogen atoms: 0.7957(32) (Mn1), 0.7711(31) (Mn2), 0.4337(32) (Mn3).

Due to disordered solvent molecules and missing reflection data, the electron density at the apical positions of the square pyramidal manganese subunits is lower than the expected values. Therefore, two plausible solutions to the diffraction data set exist, featuring three disordered chloro ligands with 2/3 occupancy or three chloro ligands with full occupancy. The first solution renders a [(Mn(II)Cl)$_2$Mn(II)Ca] core with three Mn(II) centers while the latter one features a [(Mn(II)Cl)$_2$(Mn(III)Cl)Ca] core with one Mn(III) and two Mn(II) centers. The [(Mn(II)Cl)$_2$Mn(II)Ca] core would consist of two anionic manganese(II) subunits coordinated in square pyramidal fashion and one square planar coordinated manganese(II) center. Since d^5 metal centers

103

aspire octahedral coordination geometry, square planar coordination of a manganese(II) center is disfavored and considered unlikely. As the chloride content of [16$_{Ca}$][Cl$_3$] was determined to 7.18% by ion chromatography, corresponding to three chloro ligands per molecule, the most probable solution features a [(Mn(II)Cl)$_2$(Mn(III)Cl)Ca] core. Mn1 and Mn2 are coordinated in similar fashion with τ_5 values of 0.03 and are displaced by 0.7957(32) Å (Mn1) and 0.7711(31) Å (Mn2) out of the plane produced by the coordinating oxygen and nitrogen atoms, toward the respective chloro ligands (Figure B.3.13(b) and (c)). Mn3 is coordinated in a slightly more distorted fashion with a τ_5 value of 0.10 and resides closer to the ONNO plane (0.4337(32) Å) than Mn1 and Mn2 (Figure B.3.13(d)). The different coordination geometry of Mn3 compared to Mn1 and Mn2 indicates that Mn3 may be in the formal oxidation state +III, whereas Mn1 and Mn2 are in the formal oxidation sate +II. Valence bond sum analysis supports the oxidation state assignment, rendering the molecular structure in Figure B.3.13 the most probable one (Table B.3.1). The C–O (1.303(7) – 1.334(6) Å) and C=N bond distances (1.259(8) – 1.301(7) Å) are similar to other enol-imine tautomers of tris(ONNO)-type ligands (C–O (1.328 – 1.370) and C=N (1.224 – 1.298 Å)).[48, 49]

Table B.3.1. Valence bond sum and selected coordination parameters of the three manganese centers.

atom	τ_5	V (Mn(II))[50, 51]	V (Mn(III))[50, 52]	assigned oxidation state	d(Mn – (ONNO)-plane) [Å]
Mn1	0.03	2.33	2.24	+II	0.7957(32)
Mn2	0.03	2.49	2.38	+II	0.7711(31)
Mn3	0.10	2.76	3.16	+III	0.4337(32)

Since [16$_{Ca}$][Cl$_3$] features three paramagnetic manganese centers, EPR spectroscopy was used to gain further insight into the coordination environment of the paramagnetic centers. Recording X-band EPR spectra in solution at 298 K or in frozen solution at 77 K did not reveal any resonances, due to thermal relaxation processes.

To further understand the overall properties of complex [16$_{Ca}$][Cl$_3$], the magnetic properties were determined. The effective magnetic moment μ_{eff} of [16$_{Ca}$][Cl$_3$] in solution was determined by Evans method at 298 K. The μ_{eff} value of 9.95 μ_B is lower than the expected spin-only value for three non-interacting manganese(II) high-spin centers with μ_s = 15.97 μ_B (g = 2.00).[53] To further understand the electronic configuration of the manganese centers, the magnetic data of [16$_{Ca}$][Cl$_3$] were recorded with a SQUID magnetometer. The magnetic data of [16$_{Ca}$][Cl$_3$] are presented as $\chi_m T$ as function of the temperature at 0.1 T and as molar magnetization M_m vs. the magnetic field B at 2.0 K (Figure B.3.14). The $\chi_m T$ value of 11.51 cm^3 K mol^{-1} (μ_{eff} = 9.67 μ_B) at 290 K is in the range of 10.99–12.31 cm^3 K mol^{-1} that is expected for one non-interacting Mn(III) and two non-interacting Mn(II) centers. The $\chi_m T$ values remain approximately constant in the temperature range of 100 – 290 K. At temperatures below 100 K, the values continuously decrease to 5.77 cm^3 K mol^{-1} at 2.0 K. The Mn(II) centers with 3d^5 electron configuration are described as isotropic spin centers with an effective spin of S = 5/2, whereas the distorted square-pyramidal Mn(III) center (3d^4) is slightly anisotropic. The anisotropy is attributed to a moderate split of the octahedral ^4E_g ground term over a few hundred wavenumbers resulting from the ligand field. While the drop-off of the $\chi_m T$ values below 100 K is mostly caused by predominant antiferromagnetic exchange interactions between all manganese centers, there are further contributions from the anisotropic Mn(III) center due to ligand field, spin-orbit coupling and interelectronic repulsion (single-ion effects). The molar magnetization M_m shows a linear behavior up to 1 T at 2.0 K. Above 1 T, the graph shows a curvature and the molar magnetization increases at a slower rate to a value of M_m = 9.9 N_A μ_B without being saturated. The value of the molar magnetization at 5 T is smaller than the expected saturation value of 14 N_A μ_B due to predominant antiferromagnetic exchange interactions.

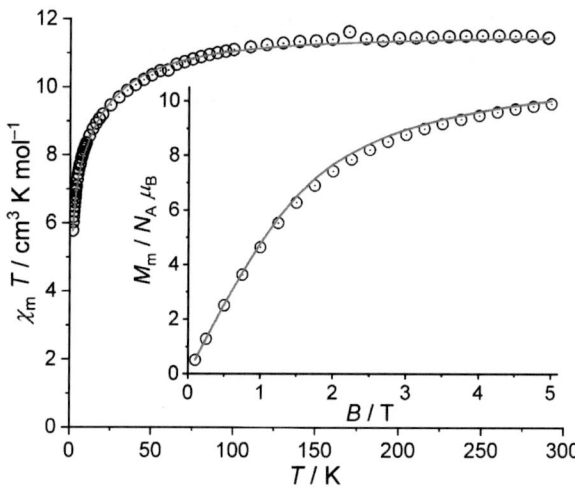

Figure B.3.14. Temperature dependence of the $\chi_m T$ values of [16$_{Ca}$][Cl$_3$] at 0.1 T; inset: molar magnetization M_m vs. applied field B at 2.0 K; experimental data (open circles), least-squares fit (solid lines) of two Mn^{2+} centers (as effective S = 5/2 centers) interacting with a single Mn^{3+} center (using the full model).

An effective isotropic spin model for the Mn(II) centers (S = 5/2, g_{eff} = 2.0) and a 'full' model of the Mn(III) center (3d^4, approximate C_{4v} symmetry) was used to quantify the magnetic properties using the CONDON software.[42, 43] For the Mn(III) center, all 210 energy states of the $3d^4$ electron configuration are considered for the single ion effects. For the Heisenberg exchange interactions, the eight lowest energy states are considered, which originate from the octahedral ^4E_g term. The single ion effects and Heisenberg exchange interactions are calculated in terms of the '$-2J$' notation. The parameters of the least-squares fit are presented in Table B.3.2. The fit with a relative mean squared error (SQ) of 1.6% affords the ligand field parameters of the Mn(III) center, describing a slightly distorted square-pyramidal ligand field that is in agreement with the molecular structure determined by X-ray diffraction analysis on single crystals (Figure B.3.13). The weak antiferromagnetic exchange interactions between the Mn(III) and one of the Mn(II) centers and between the two Mn(II) centers are described by the exchange interaction parameters that are given by $J_1 = -0.34$ cm^{-1} (Mn(II)–Mn(III)) and $J_2 = -0.09$ cm^{-1} (Mn(II)–Mn(II)). The small magnitude of the antiferromagnetic exchange interactions is due to the large distance between the manganese centers and due to the potential alternation of the oxidation states among the Mn centers. Unlike the calcium trivanadyl complex [13$_{Ca}$][(OTf)$_2$], the calcium trimanganese complex [16$_{Ca}$][Cl$_3$] cannot be described as a frustrated spin system as the spins per manganese center are larger than 1/2, allowing further coupling scenarios. The magnetic data of [16$_{Ca}$][Cl$_3$] substantiates the formation of a trimanganese complex with one Mn(III) and two Mn(II) centers.

Table B.3.2. Parameters of the least-squares fit of the SQUID data of [16$_{Ca}$][Cl$_3$] with [a] one electron spin-orbit coupling constant[54], [b] Racah parameters[54], ligand field B^k_q parameters in Wybourne notation, and exchange interaction parameters in '$-2J$' notation (J_1: Mn(II)–Mn(III), J_2: Mn(II)–Mn(II)).

	Mn(II)	Mn(III)
g_{eff}	2.0	/
S_{eff}	5/2	/
ζ_{3d} [cm^{-1}][a]	/	352
B [cm^{-1}][b]	/	1140
C [cm^{-1}][b]	/	3675
B^2_0 [cm^{-1}]	/	−2144 ± 76
B^4_0 [cm^{-1}]	/	46812 ± 978
B^4_4 [cm^{-1}]	/	25013 ± 210
J_1 [cm^{-1}]	−0.34 ± 0.01	
J_2 [cm^{-1}]	−0.09 ± 0.01	
SQ [%]	1.6	

To probe whether [16$_{Ca}$][Cl$_3$] can be reversibly oxidized or reduced, the electrochemical properties of the complex were determined by cyclic voltammetry (Figure B.3.15). The cyclic voltammogram of [16$_{Ca}$][Cl$_3$] in DCM reveals an irreversible oxidation event at a potential $E_p = 0.99$ V with a preceding broad, irreversible oxidation in the range of −0.22 to 0.80 V (vs Fc/Fc$^+$, compound (2 mM), electrolyte: [^{n}Bu$_4$N][PF$_6$] (100 mM)). A broad reduction event is observed at negative potentials in the range of −0.89 to −2.02 V. To further understand the redox events, cyclic voltammograms at a higher scan rate of 500 mV s^{-1} were recorded (Figure B.3.15(b)). After completion of the first cycle, the peak current of the oxidation event at E_p = 1.12 V remains approximately constant. An additional reduction event is observed at E_p = −0.43 V which is not observed at a scan rate of 250 mV s^{-1}. After completion of the first cycle, the peak current of the reduction event also remains approximately constant but is smaller than for the oxidation event at E_p = 1.12 V. Presumably, [16$_{Ca}$][Cl$_3$] is irreversibly oxidized at E_p = 1.12 V to produce a new species in a subsequent reaction. At a lower scan rate and, thus, at longer measurement times, the generated species is not observed, due to complete decomposition during the measurement. When recording five cyclic voltammograms of the oxidation event at E_p = 1.12 V in a smaller measurement window, the initial species is gradually consumed and the peak current decreases with each cycle (Figure B.3.15(c)). Once the measurement window is expanded again to include the reduction event at E_p = −0.43 V, the peak current increases and stays approximately constant with each cycle. Due to the low resolution of the cyclic voltammograms, the redox events cannot be undoubtedly assigned to metal- or ligand-centered processes.

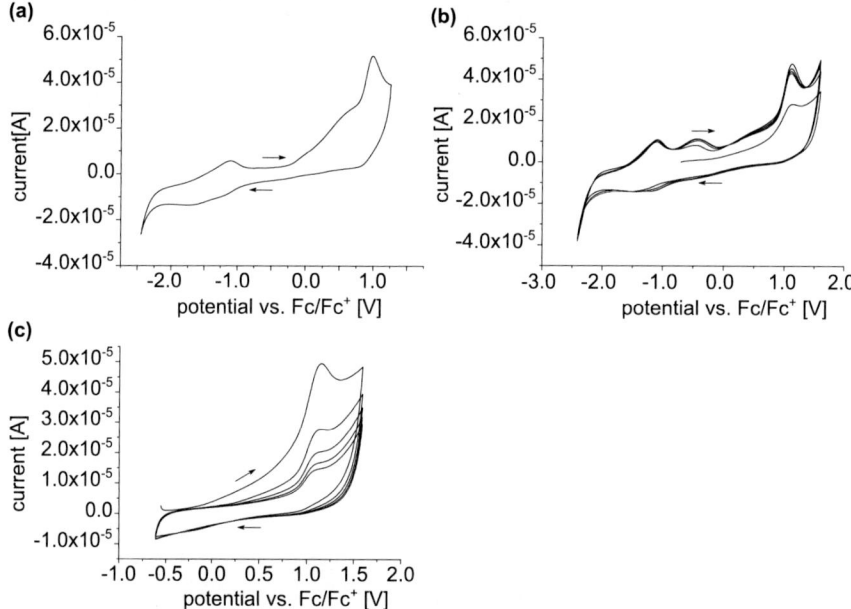

Figure B.3.15. Cyclic voltammograms of [16$_{Ca}$][Cl$_3$] (2 mM) in DCM at 298 K, electrolyte [^{n}Bu$_4$N][PF$_6$] (100 mM), referenced to the Fc/Fc$^+$ couple; (a) complete cycle recorded at a scan rate of 250 mV s^{-1}; (b) five cycles recorded at a scan rate of 500 mV s^{-1}; (c) five cycles of the oxidation event recorded at a scan rate of 500 mV s^{-1}.

B.3.3 Summary and Outlook

In this chapter a new approach toward heterometallic complexes featuring tris(ONNO)-type macrocyclic ligands based on 3,6-diformylcatechol was established. The complexes are synthesized in a two-stage reaction by first preparing the mononuclear complexes and by subsequent metalation of the salen-type binding sites (Scheme B.3.10). This approach resolves the drawbacks of previous methods relying on heterometallic template directed synthesis or metalation of protonated pro-ligands. The first method is limited to certain metal combinations,[19-26] whereas the latter is limited to pro-ligands featuring rigid benzene-1,2-diamine bridging units that decrease the solubility of the complexes due to π-π interactions.[4, 49, 55] Additionally, the synthesis of pro-ligands is prone to undesired side reactions.[27, 56]

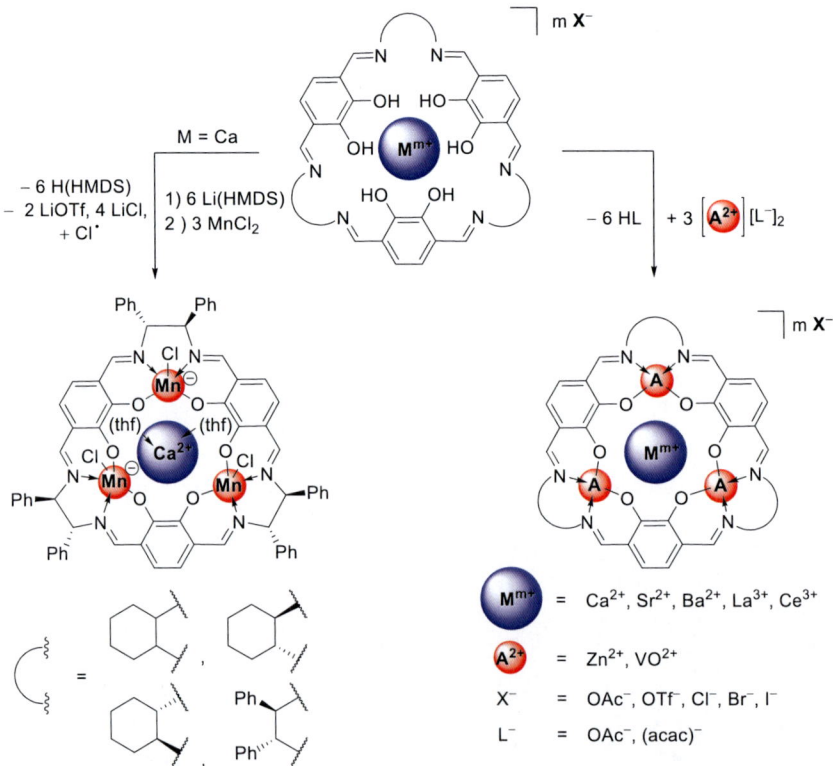

Scheme B.3.10. Preparation of heterometallic tetranuclear complexes form mononuclear starting complexes, stabilized by a central cation in the 18c6-type cavity.

The alkaline earth metal or lanthanide metal cation in the central 18c6-type cavity sufficiently stabilizes the tris(ONNO)-type ligand framework against reversible Schiff base rearrangements

Results and Discussion

as was shown with the trizinc system. The tetranuclear calcium trizinc complex [11$_{Ca}$][I$_2$] was selectively prepared by heterometallic template directed synthesis or by the two-stage synthesis. In the latter method, the mononuclear complex [9$_{Ca}$][I$_2$] was first prepared and, subsequently, the three salen (ONNO)-type binding sites were metalated with zinc cations. [11$_{Ca}$][(OAc)$_2$] and [11$_{Ce}$][Cl$_3$] were prepared similar to [11$_{Ca}$][I$_2$] by heterometallic template synthesis. The complexes were characterized by mass spectrometry, NMR and IR spectroscopy. The molecular structure of [11$_{Ca}$][(OAc)$_2$] confirms exclusive location of the zinc cations in the salen-type binding sites, whereas calcium is located in the 18c6-type cavity. A VT NMR study on [11$_{Ca}$][(OAc)$_2$] in DMSO-d_6 indicates an exchange process due to conformational changes of the macrocyclic ligand backbone or an exchange between coordinating and non-coordinating acetate.

Trivanadyl and trimanganese complexes were synthesized form the mononuclear complexes by the two-stage approach. The calcium trivanadyl complex [12$_{Ca}$][(OTf)$_2$] featuring the (R,R)-1,2-cyclohexanediamine backbone was prepared from [7$_{Ca}$][(OTf)$_2$] and VO(acac)$_2$. The complex was analyzed by IR spectroscopy and mass spectrometry. Due to the low solubility of [12$_{Ca}$][(OTf)$_2$], calcium trivanadyl complexes [13$_{Ca}$][(OTf)$_2$], [13$_{Ca}$][Cl$_2$], [13$_{Ca}$][Br$_2$] and [13$_{Ca}$][I$_2$] featuring a (R,R)-1,2-diphenylethylenediamine bridging unit were synthesized. All complexes were characterized by mass spectrometry, EPR and IR spectroscopy. [13$_{Ca}$][(OTf)$_2$] is another example of a frustrated spin system with predominantly antiferromagnetic exchange interactions at low temperatures. The tetranuclear complexes [13$_{Sr}$][(OTf)$_2$], [13$_{Ba}$][(OTs)$_2$], [13$_{La}$][(OTf)$_3$] and [13$_{Ce}$][Cl$_3$] featuring strontium, barium, lanthanum and cerium(III) in the central 18c6-type cavity were synthesized and characterized similar to [13$_{Ca}$][(OTf)$_2$]. All trivanadyl complexes exhibit similar properties and only subtle changes are observed by spectroscopic means: Whereas most of the prepared trivanadyl complexes produce slightly anisotropic EPR signals with clearly distinguishable hyperfine coupling patterns, the complexes [13$_{Ca}$][Br$_2$], [13$_{Ca}$][I$_2$] and [13$_{Sr}$][(OTf)$_2$] produce complex EPR spectra, due to presumed inequivalence of the vanadium centers.

The tetranuclear calcium trimanganese complex [16$_{Ca}$][Cl$_3$] was synthesized through deprotonation of the mononuclear complex [9$_{Ca}$][(OTf)$_2$] with LiHMDS and subsequent salt metathesis with MnCl$_2$. The complex features one manganese(III) and two manganese(II) centers, and was characterized by IR spectroscopy, mass spectrometry, X-ray diffraction and magnetic analysis. Similar to the calcium trivanadyl complex [13$_{Ca}$][(OTf)$_2$], the spins of the manganese centers of [16$_{Ca}$][Cl$_3$] are coupled to produce predominantly antiferromagnetic exchange interactions. The complex features a [Mn$_3$Ca] core which may allow structural and comparative studies to the [Mn$_3$Ca] cubane found in PS II.

B.3.4 Experimental

B.3.4.1 General Considerations

1,4-diformyl-2,3-dimethoxybenzene,[25, 57] 3,6-diformylcatechol[25, 57] were prepared according to literature procedures. The monometallic complexes were synthesized according to procedures reported in chapter B.2. VO(acac)$_2$ was purified through recrystallization from DCM. All other chemicals were used without further purification. Monometallic and tetranuclear trizinc complexes were synthesized under atmospheric conditions without dried solvents. Trivanadyl complexes were synthesized with degassed solvents. trimanganese complexes were prepared with dried and degassed solvents.

B.3.4.2 Synthesis of Trizinc Complexes

[((R,R)-^{C2Ph2}L)Zn$_3$CaI$_2$] ([11$_{Ca}$][I$_2$])

Method A:

An orange solution of Zn(OAc)$_2$ · 2 H$_2$O (66.1 mg, 301 μmol, 3.00 equiv.), CaI$_2$ · H$_2$O (31.3 mg, 100 μmol, 1.00 equiv.) and 3,6-diformylcatechol (50.0 mg, 301 μmol, 3.00 equiv.) in MeCN/MeOH (1:1, 20 mL) were heated under reflux for 30 min. (R,R)-(+)-1,2-diphenylethylenediamine (63.9 mg, 301 μmol, 3.00 equiv.) in MeOH/MeCN (1:1, 4 mL) was added to the hot solution. The solution was heated for an additional 16 h under reflux to produce an orange suspension. The solid was collected and washed with Et$_2$O. Drying under reduced pressure afforded **[11$_{Ca}$][I$_2$]** as an orange solid (144 mg, 95.8 μmol, 96%).

Method B:

Zn(OAc)$_2$ · 2 H$_2$O (34.7 mg, 158 μmol, 3.00 equiv.) in MeCN/MeOH (5 mL) was added to a red solution of **[9$_{Ca}$][I$_2$]** (69.6 mg, 52.7 μmol, 1.00 equiv.) in MeCN/MeOH (1:1, 10 mL). The red solution was heated under reflux for 16 h to produce an orange suspension. The orange solid was collected and washed with MeOH. Drying under reduced pressure afforded **[11$_{Ca}$][I$_2$]** as an orange solid (41.9 mg, 27.7 μmol, 53%). ^{1}**H NMR** (DMSO-d_6): δ 8.01 (s, 6H, HC=N), 7.42 – 7.25 (m, 30H, H$_{aryl}$), 6.20 (s, 6H, H$_{aryl}$), 5.27 (br, 6H, CH). **FAB MS** (pos. mode, 3-NBA matrix): m/z = 1420.9 ([((R,R)-^{C2Ph2}L)Zn$_3$Ca + 3-NBA]$^+$). **IR** in KBr: ($\tilde{v}$, cm^{-1}) 3440 (m), 3028 (w), 2894 (w), 1622 (vs, C=N), 1582 (m), 1460 (s), 1404 (w), 1316 (m), 1233 (m), 1165 (m), 1026 (w), 970 (w), 869 (w), 835 (w), 759 (w), 736 (m), 701 (m), 667 (vw), 608 (vw), 558 (vw), 467 (vw). **Elemental Analysis:** Calcd. for C$_{66}$H$_{48}$CaI$_2$N$_6$O$_6$Zn$_3$(Et$_2$O): C 53.04, H 3.69, N 5.30; found: C 53.42, H 3.68, N 5.28.

[((R,R)-^{C2Ph2}L)Zn$_3$Ca(OAc)$_2$] ([11$_{Ca}$][(OAc)$_2$])

A mixture of Ca(OAc)$_2$ · 2 H$_2$O (19.5 mg, 100 μmol, 1.00 equiv.) and 3,6-diformylcatechol (50.0 mg, 300 μmol, 3.00 equiv.) and Zn(OAc)$_2$ · 2 H$_2$O (66.1 mg, 300 μmol, 3.00 equiv.) in

Results and Discussion

MeOH/MeCN (1:1, 15 mL) was heated under reflux for 15 min to produce an orange solution. (R,R)-(+)-1,2-diphenylethylenediamine (63.9 mg, 300 μmol, 3.00 equiv.) in MeOH/MeCN (1:1, 4 mL) was added to the hot solution. The solution was heated under reflux for an additional 3 h to produce an orange suspension. The orange solid was collected through centrifugation and rinsed with Et$_2$O. Drying under reduced pressure afforded **[11$_{Ca}$][(OAc)$_2$]** as an orange solid (77.0 mg, 56.0 μmol, 56%). ^{1}**H NMR** (DMSO-d_6): δ 7.92 (s, 6H, HC=N), 7.52 – 7.09 (m, 30H, H_{aryl}), 6.05 (s, 6H, H_{aryl}), 5.36 – 4.80 (br, 6H, CH), 1.86 – 1.52 (br, 6H, CH$_3$). **FAB MS** (pos. mode, 3-NBA matrix): m/z = 1313.1 ([((R,R)-^{C2Ph2}L)Zn$_3$Ca(OAc)]$^+$). **IR** in KBr: ($\tilde{v}$, cm^{-1}) 3424 (br. w), 3028 (w), 2895 (w), 1616 (vs, C=N), 1578 (m), 1490 (m), 1459 (s), 1398 (m), 1327 (m), 1213 (w), 1163 (m), 1119 (vw), 1029 (vw), 1002 (vw), 970 (w), 918 (vw), 865 (vw), 835 (w), 760 (w), 737 (m), 703 (m), 680 (vw), 613 (w), 561 (w), 536 (w), 499 (vw). **Elemental Analysis:** Calcd. for C$_{70}$H$_{54}$CaN$_6$O$_{10}$Zn$_3$(H$_2$O)$_2$: C 59.57, H 4.14, N 5.95; found: C 59.33, H 4.15, N 5.76. **XRD:** Yellow crystals suitable for single crystal X-ray diffraction analysis were obtained through slow evaporation of a solution of **[11$_{Ca}$][(OAc)$_2$]** in DMSO-d_6.

[((R,R)-^{C2Ph2}L)Zn$_3$CeCl$_3$] ([11$_{Ce}$][Cl$_3$])

A mixture of CeCl$_3$ (12.4 mg, 50.0 μmol, 1.00 equiv.) and 3,6-diformylcatechol (25.0 mg, 150 μmol, 3.00 equiv.) and Zn(OAc)$_2$ · 2 H$_2$O (33.0 mg, 150 μmol, 3.00 equiv.) in MeOH/MeCN (1:1, 15 mL) was heated under reflux for 15 min to produce an orange solution. (R,R)-(+)-1,2-diphenylethylenediamine (39.9 mg, 150 μmol, 3.00 equiv.) in MeOH/MeCN (1:1, 4 mL) was added to the hot solution. The solution was heated under reflux for an additional 3 h to produce an orange suspension. The orange solid was collected through centrifugation and rinsed with Et$_2$O. Drying under reduced pressure afforded **[11Ce][Cl3]** as an orange solid (77.0 mg, 56.0 μmol, 56%). ^{1}**H NMR** (DMSO-d_6): δ 8.93 – 8.30 (br d, 6H, HC=N), 7.85 – 6.86 (m, 30H, H_{aryl}), 6.18 – 5.49 (s, 6H, H_{aryl}), 4.03 (br, 3H, CH$_a$), 1.93 (br, 3H, CH$_b$). **FAB MS** (pos. mode, 3-NBA matrix): m/z = 1428.0 ([((R,R)-^{C2Ph2}L)Zn$_3$CeCl$_2$]$^+$). **IR** in KBr: ($\tilde{v}$, cm^{-1}) 3424 (m), 3028 (w), 2894 (w), 1619 (vs, C=N), 1584 (w), 1509 (m), 1495 (m), 1463 (s), 1447 (s), 1390 (m), 1320 (s), 1233 (s), 1166 (m), 1076 (vw), 1028 (vw), 988 (w), 969 (w), 865 (w), 835 (w), 762 (w), 738 (m), 701 (m), 682 (w), 611 (w), 563 (w), 535 (w), 504 (w). **Elemental Analysis:** Calcd. for C$_{66}$H$_{48}$CeCl$_3$N$_6$O$_6$Zn$_3$(H$_3$COH)$_2$: C 53.46, H 3.69, N 5.50; found: C 52.63, H 3.99, N 5.61.

B.3.4.3 Synthesis of Trivanadyl Complexes

[((R,R)-^{Cy}L)(VO)$_3$Ca(OTf)$_2$] ([12$_{Ca}$][(OTf)$_2$])

A solution of VO(acac)$_2$ (18.6 mg, 70.0 μmol, 3.00 equiv.) in MeOH (5 mL) was added to a stirred suspension of **[7$_{Ca}$][(OTf)$_2$]** (25.0 mg, 23.3 μmol, 1.00 equiv.) in MeOH (7 mL). The reaction mixture was stirred for 16 h at r.t. to produce a dark brown solution. All volatiles were removed under reduced pressure and the dark orange/brown solid rinsed with Et$_2$O and n-

Heterometallic Tetranuclear Complexes Featuring a Tris(ONNO)-Type Ligand

hexane. The product [12$_{Ca}$][(OTf)$_2$] was obtained as a dark orange/brown solid (29.2 mg, 23.0 μmol, 99%). **FAB MS** (pos. mode, 3-NBA matrix): m/z = 1116.0 ([^{Cy}L(VO)$_3$Ca(OTf)]$^+$), 967.1 ([((R,R)-^{Cy}L)(VO)$_3$Ca + e⁻]$^+$), 901.1 ([H((R,R)-^{Cy}L)(VO)$_2$Ca]$^+$), 1051.1 [(H$_2$((R,R)-^{Cy}L)(VO)$_2$Ca(OTf)]$^+$). **IR** in KBr: ($\tilde{v}$, cm^{-1}) 3435 (s), 2940 (m), 1630 (vs, C=N), 1524 (s), 1468 (m) 1448 (m), 1384 (m), 1328 (vs), 1280 (s), 1251 (s), 1169 (m), 1031 (s), 996 (w, V=O), 894 (vw), 865 (vw), 851 (vw), 789 (vw), 765 (w), 671 (w), 639 (m), 607 (w), 518 (w).

[((R,R)-^{C2Ph2}L)(VO)$_3$Ca(OTf)$_2$] ([13$_{Ca}$][(OTf)$_2$])

A solution of VO(acac)$_2$ (58.3 mg, 220 μmol, 3.00 equiv.) and [9$_{Ca}$][(OTf)$_2$] (100 mg, 73.3 μmol, 1.00 equiv.) in MeCN/MeOH (1:1) was stirred for 20 h at r.t. to produce a dark brown solution. All volatiles were removed under reduced pressure. The brown solid was extracted with DCM and filtered. An orange precipitate was produced through addition of Et$_2$O to the solution. The orange solid was collected, taken up in DCM and again precipitated. The solid was collected and dried under reduced pressure which afforded [13$_{Ca}$][(OTf)$_2$] as an orange/brown, air sensitive solid (95.1 mg, 61.0 μmol, 83%). **FAB MS** (pos. mode, 3-NBA matrix): m/z = 1260.0 ([((R,R)-^{C2Ph2}L)(VO)$_3$Ca]$^+$), 1425.9 ([((R,R)-^{C2Ph2}L)(VO)$_3$Ca(OTf) + MeOH]$^+$), 1343.9 ([((R,R)-^{C2Ph2}L)(VO)$_3$Ca + 2 MeCN]$^+$), 1360.9 ([((R,R)-^{C2Ph2}L)(VO)$_3$Ca(OH) + 2MeCN]$^+$), 1408.8 ([((R,R)-^{C2Ph2}L^{C2Ph2}L)(VO)$_3$Ca(OTf) + 2MeOH + 2MeCN]$^+$). **IR** in KBr: ($\tilde{v}$, cm^{-1}) 3440 (m), 3064 (w), 3030 (w), 2922 (w), 1616 (vs, C=N), 1587 (m), 1514 (m), 1497 (m), 1467 (m), 1454 (s), 1384 (m), 1327 (s), 1249 (s), 1165 (m), 1030 (s), 1000 (m, V=O), 867 (w), 839 (w), 756 (m), 739 (m), 702 (m), 638 (s), 607 (vw), 572 (w), 545 (w), 517 (w), 477 (w). **Elemental Analysis:** Calcd. for C$_{68}$H$_{48}$CaF$_6$N$_6$O$_{15}$S$_2$V$_3$: C 52.35, H 3.10, N 5.39; found: C 54.30, H 3.97, N 5.71. **EPR** (10 mM in MeCN, frequency: 9.410 GHz, modulation amplitude: 0.200 mT, attenuation: 10 dB, sweep time: 60 s; B_0 sweep: 200 mT, B_0: 340 mT): g_{iso} = 1.96, A_{iso} = 292 MHz. **Magnetic Properties**: $\chi_m T$ = 1.09 cm^3 K mol^{-1}, μ_{eff} = 2.91 μ_B (SQUID, 290 K).

[((R,R)-^{C2Ph2}L)(VO)$_3$CaCl$_2$] ([13$_{Ca}$][Cl$_2$])

[13$_{Ca}$][Cl$_2$] was prepared analogously to [13$_{Ca}$][(OTf)$_2$] with [9$_{Ca}$][Cl$_2$] (125 mg, 109 μmol, 1.00 equiv.) and VO(acac)$_2$ (87.1 mg, 328 μmol, 3.00 equiv.) in MeOH/MeCN (1:1). The product [13$_{Ca}$][Cl$_2$] was obtained as a dark orange/red solid (74.1 mg, 55.6 μmol, 51%). **FAB MS** (pos. mode, 3-NBA matrix): m/z = 1296.0 ([((R,R)-^{C2Ph2}L)(VO)$_3$CaCl]$^+$), 1194.1 ([H((R,R)-^{C2Ph2}L)(VO)$_2$Ca]$^+$), 1230.0 ([H$_2$((R,R)-^{C2Ph2}L)(VO)$_2$CaCl]$^+$), 1129.2 ([H$_3$((R,R)-^{C2Ph2}L)(VO)Ca]$^+$). **IR** in KBr: ($\tilde{v}$, cm^{-1}) 3416 (w), 3028 (w), 2918 (w), 1616 (vs, C=N), 1537 (m), 1498 (m), 1451 (s), 1384 (vs), 1327 (s), 1230 (s), 1182 (m), 1120 (m), 1054 (w), 1001 (s, V=O), 787 (m), 700 (m), 646 (w), 575 (w), 546 (w), 485 (w). **EPR** (10 mM in MeCN, frequency: 9.421 GHz,

Results and Discussion

modulation amplitude: 0.200 mT, attenuation: 10 dB, sweep time: 60 s; B_0 sweep: 200 mT, B_0: 340 mT): g_{eff} = 1.96, A = 292 MHz.

[((R,R)-^{C2Ph2}L)(VO)$_3$CaBr$_2$] ([13$_{Ca}$][Br$_2$])

[13$_{Ca}$][Br$_2$] was prepared analogously to [13$_{Ca}$][(OTf)$_2$] with [9$_{Ca}$][Br$_2$] (121 mg, 98.9 µmol, 1.00 equiv.) and VO(acac)$_2$ (78.6 mg, 297 µmol, 3.00 equiv.) in MeOH/MeCN (1:1). The product [13$_{Ca}$][Br$_2$] was obtained as a dark red/orange solid (135 mg, 94.9 µmol, 96%). **FAB MS** (pos. mode, 3-NBA matrix): m/z = 1194.1 ([H((R,R)-^{C2Ph2}L)(VO)$_2$Ca]$^+$), 1129.2 ([H$_3$((R,R)-^{C2Ph2}L)(VO)Ca]$^+$), 1425.8 ([((R,R)-^{C2Ph2}L)(VO)$_3$CaBr$_2$]$^+$). **IR** in KBr: ($\tilde{v}$, cm^{-1}) 3028 (vw), 2877 (vw), 1616 (vs, C=N), 1538 (m), 1496 (m), 1451 (s), 1384 (m), 1328 (vs), 1228 (m), 1182 (m), 1119 (w), 1031 (m), 1002 (s, C=N), 884 (w), 830 (w), 780 (m), 700 (s), 645 (w), 575 (w), 546 (w), 477 (w). **Elemental Analysis** Calcd. for C$_{66}$H$_{48}$CaBr$_2$N$_6$O$_9$V$_3$(Et$_2$O): C 59.75, H 4.16, N 5.97; found: C 59.65, H 4.65, N 6.06.

[((R,R)-^{C2Ph2}L)(VO)$_3$CaI$_2$] ([13$_{Ca}$][I$_2$])

[13$_{Ca}$][I$_2$] was prepared analogously to [13$_{Ca}$][(OTf)$_2$] with [9$_{Ca}$][I$_2$] (198 mg, 150 µmol, 1.00 equiv.) and VO(acac)$_2$ (119 mg, 450 µmol, 3.00 equiv.) in MeOH/MeCN (1:1). The product [13$_{Ca}$][I$_2$] was obtained as a dark red/orange solid (182 mg, 120 µmol, 80%). **FAB MS** (pos. mode, 3-NBA matrix): m/z = 1515.0 ([((R,R)-^{C2Ph2}L)(VO)$_3$CaI$_2$]$^+$), 1386.9 ([((R,R)-^{C2Ph2}L)(VO)$_3$CaI]$^+$), 1260.1 ([((R,R)-^{C2Ph2}L)(VO)$_3$Ca]$^+$), 1194.2 ([((R,R)-^{C2Ph2}L)(VO)$_2$Ca]$^+$), 1426.0 ([((R,R)-^{C2Ph2}L)(VO)$_3$Ca + 3-NBA + H]$^+$). **IR** in KBr: ($\tilde{v}$, cm^{-1}) 3423 (m), 3029 (w), 2919 (w), 1614 (vs, C=N), 1540 (w), 1509 (m), 1467 (m), 1449 (s), 1384 (s), 1327 (vs), 1233 (s), 1184 (m), 999 (m, V=O), 866 (vw), 839 (m), 755 (m), 702 (m), 645 (w), 684 (vw), 571 (vw), 546 (vw), 476 (vw). **Elemental Analysis:** Calcd. for C$_{66}$H$_{48}$CaI$_2$N$_6$O$_9$V$_3$(Et$_2$O)$_3$: C 53.90, H 4.52, N 4.83; found: C 54.75, H 4.53, N 5.14.

[((R,R)-^{C2Ph2}L)(VO)$_3$Sr(OTf)$_2$] ([13$_{Sr}$][(OTf)$_2$])

[13$_{Sr}$][(OTf)$_2$] was prepared analogously to [13$_{Ca}$][(OTf)$_2$] with [9$_{Sr}$][(OTf)$_2$] (141 mg, 100 µmol, 1.00 equiv.) and VO(acac)$_2$ (79.5 mg, 300 µmol, 3.00 equiv.) in MeOH/MeCN (1:1). The product [13$_{Sr}$][(OTf)$_2$] was obtained as a dark red/orange solid (89.0 mg, 55.0 µmol, 55%). **FAB MS** (pos. mode, 3-NBA matrix): m/z = 1456.6 ([((R,R)-^{C2Ph2}L)(VO)$_3$Sr(OTf)]$^+$), 1391.7 ([H$_2$((R,R)-^{C2Ph2}L)(VO)$_2$Sr(OTf)]$^+$), 1241.9 (H[((R,R)-^{C2Ph2}L)(VO)$_2$Sr]$^+$). **IR** in KBr: ($\tilde{v}$, cm^{-1}) 3441 (w), 3061 (w), 3030 (w), 2921 (w), 1614 (vs, C=N), 1540 (m), 1498 (m), 1465 (s), 1454 (s),

1384 (s), 1326 (s), 1260 (vs), 1227 (s), 1172 (m), 1120 (w), 1030 (s), 1003 (m, V=O), 867 (vw), 830 (vw), 760 (m), 758 (m), 701 (m), 638 (s), 574 (w), 546 (vw), 517 (w), 482 (vw).

[((R,R)-^{C2Ph2}L)(VO)$_3$Ba(OTs)$_2$] ([13$_{Ba}$][(OTs)$_2$])

[13$_{Ba}$][(OTs)$_2$] was prepared analogously to [13$_{Ca}$][(OTf)$_2$] with [9$_{Ba}$][(OTs)$_2$] (151 mg, 100 μmol, 1.00 equiv.) and VO(acac)$_2$ (79.5 mg, 300 μmol, 3.00 equiv.) in MeOH/MeCN (1:1). The product [13$_{Ba}$][(OTs)$_2$] was obtained as a dark red/orange solid (141 mg, 83 μmol, 83%). **FAB MS** (pos. mode, 3-NBA matrix): m/z = 1529.4 ([((R,R)-^{C2Ph2}L)(VO)$_3$Ba(OTs)]$^+$), 1292.6 ([H((R,R)-^{C2Ph2}L)(VO)$_2$Ba]$^+$). **IR** in KBr: ($\tilde{v}$, cm^{-1}) 3422 (w), 3056 (w), 3029 (w), 2920 (w), 1614 (vs, C=N), 1584 (m), 1530 (s), 1496 (w), 1452 (w), 1384 (s), 1323 (s), 1230 (vs), 1176 (s), 1122 (s), 1033 (s), 1010 (s), 1001 (s, V=O), 864 (vw), 815 (w), 786 (m), 702 (m), 681 (s), 640 (vw), 619 (vw), 569 (m), 485 (w). **EPR** (10 mM in MeCN, frequency: 9.420 GHz, modulation amplitude: 0.200 mT, attenuation: 10 dB, sweep time: 60 s; B_0 sweep: 200 mT, B_0: 340 mT): g_{eff} = 1.96, A = 292 MHz.

[((R,R)-^{C2Ph2}L)(VO)$_3$La(OTf)$_3$] ([13$_{La}$][(OTf)$_3$])

[13$_{La}$][(OTf)$_3$] was prepared analogously to [13$_{Ca}$][(OTf)$_2$] with [9$_{La}$][(OTf)$_3$] (150 mg, 93.0 μmol, 1.00 equiv.) and VO(acac)$_2$ (74.0 mg, 279 μmol, 3.00 equiv.) in MeOH/MeCN (1:1). The product [13$_{La}$][(OTf)$_3$] was obtained as a dark red/orange solid (157 mg, 86.8 μmol, 94%). **FAB MS** (pos. mode, 3-NBA matrix): m/z = 1656.3 ([((R,R)-^{C2Ph2}L)(VO)$_3$La(OTf)$_2$]$^+$), 1526.5 ([((R,R)-^{C2Ph2}L)(VO)$_3$La + 3-NBA]$^+$), 1441.7 ([H((R,R)-^{C2Ph2}L)(VO)$_2$La(OTf)]$^+$), 1591.4 ([H$_2$((R,R)-^{C2Ph2}L)(VO)$_2$La(OTf)$_2$]$^+$). **IR** in KBr: ($\tilde{v}$, cm^{-1}) 3423 (w), 3062 (w), 3032 (w), 2972 (w), 2874 (w), 1622 (vs, C=N), 1547 (m), 1478 (m), 1454 (m), 1384 (s), 1309 (s), 1276 (s), 1224 (vs), 1172 (m), 1029 (vs), 1003 (m, V=O), 917 (w), 837 (w), 730 (m), 701 (m), 637 (s), 572 (w), 545 (w), 516 (w), 486 (w). **EPR** (10 mM in MeCN, frequency: 9.427 GHz, modulation amplitude: 0.200 mT, attenuation: 10 dB, sweep time: 60 s; B_0 sweep: 200 mT, B_0: 340 mT): g_{eff} = 1.96, A = 292 MHz.

[((R,R)-^{C2Ph2}L)(VO)$_3$CeCl$_3$] ([13$_{Ce}$][Cl$_3$])

[13$_{Ce}$][Cl$_3$] was prepared analogously to [13$_{Ca}$][(OTf)$_2$] with [9$_{Ce}$][Cl$_3$] (100 mg, 78.5 μmol, 1.00 equiv.) and VO(acac)$_2$ (64.5 mg, 243 μmol, 3.00 equiv.) in MeOH/MeCN (1:1). The product [13$_{Ce}$][Cl$_3$] was obtained as a dark red/orange solid (89.8 mg, 61.3 μmol, 78%). **FAB MS** (pos. mode, 3-NBA matrix): m/z = 1461.2 ([((R,R)-^{C2Ph2}L)(VO)$_3$CeCl$_3$]$^+$), 1331.1 ([(H(R,R)-^{C2Ph2}L)(VO)$_2$CeCl]$^+$). **IR** in KBr: ($\tilde{v}$, cm^{-1}) 3405 (vw), 3028 (vw), 2919 (vw), 1615 (vs, C=N), 1545 (w), 1510 (w), 1468 (m), 1449 (m), 1391 (m), 1322 (s), 1230 (vs), 1181 (m), 1003 (m, V=O), 836 (vw), 730 (m), 700 (m), 642 (w), 601 (vw), 570 (vw), 545 (vw), 482 (vw). **Elemental Analysis**: Calcd. for $C_{66}H_{48}CeCl_3N_6O_9V_3(Et_2O)$: C 54.50, H 3.79, N 5.45; found: C 54.30, H 4.11, N 5.57.

B.3.4.4 Synthesis of Trimanganese Complexes

[((R,R)-C2Ph2L)(Mn(II)Cl)$_2$(Mn(III)Cl)Ca(thf)$_2$] ([16$_{Ca}$][Cl$_3$])

A solution of LiHMDS (73.5 mg, 439 μmol, 6.00 equiv) in THF (2 mL) was added to a stirred red suspension of **[9$_{Ca}$][(OTf)$_2$]** (100 mg, 73.2 μmol, 1.00 equiv) in MeCN (5 mL) upon which an orange solution was obtained. After stirring for 15 min at r.t., the solution was added to a suspension of $MnCl_2$ (27.6 mg, 220 μmol, 3.00 equiv) in THF (5 mL). Stirring was continued for 19 h at r.t. to produce a red/brown reaction mixture. All volatiles were removed under reduced pressure. The red/brown solid was extracted with DCM (2 x 5 mL) and filtered to remove LiCl and LiOTf. An orange solid was precipitated through fast addition of Et_2O to the dark orange solution in DCM. The orange solid was collected and dried under reduced pressure. The product **[16$_{Ca}$][Cl$_3$]** was obtained as an orange solid (57.2 mg, 44.1 μmol, 86%). The compound is paramagnetic and ^{1}H NMR resonances were not observed. **FAB MS** (pos. mode, 3-NBA matrix): m/z = 1260.1 ([((R,R)-C2Ph2L)Mn$_3$CaCl]$^+$), 1295.0 ([((R,R)-C2Ph2L)Mn$_3$CaCl$_2$]$^+$). **ESI MS** (pos. mode, THF): m/z calcd. for ([$C_{66}H_{51}CaCl_2Mn_3N_6O_7$]$^+$, [((R,R)-C2Ph2L)Mn$_3$CaCl$_2$ + H$_3$O$^+$]): 1315.0997; found: 1315.06958. **IR** in KBr: ($\tilde{v}$, cm^{-1}) 3422 (w), 3030 (w), 2895 (w), 1616 (vs, C=N), 1496 (w), 1453 (s), 1317 (s), 1233 (m), 1173 (w), 1031 (w), 968 (w), 918 (vw), 836 (w), 746 (w), 702 (m), 639 (w), 605 (vw). **Elemental Analysis:** Calcd. for $C_{74}H_{64}CaCl_3Mn_3N_6O_8$: C 60.19, H 4.37, N 5.69; found: C 57.85, H 4.49, N 7.27. **Chloride Content**: Calcd. for $C_{66}H_{48}CaCl_3Mn_3N_6O_6(C_4H_8O)_2$: Cl 7.20%; found: 7.18% ± 0.06%. **Manganese and Calcium Content**: Calcd. for $C_{66}H_{48}CaCl_3Mn_3N_6O_6(C_4H_8O)_2$: Mn 11.44%, Ca 2.78%; found: Mn 12.89%, Ca 1.91%. **XRD**: Dark brown/black crystals suitable for single crystal X-ray diffraction analysis were obtained through gas phase diffusion of Et_2O into a solution of **[16$_{Ca}$][Cl$_3$]** in MeCN/THF. **Magnetic Properties**: μ_{eff} = 9.95 μ_B (Evans Method, DCM); $\chi_m T$ 11.51 cm^3 K mol^{-1}, μ_{eff} = 9.67 μ_B (SQUID, 290 K).

B.3.4.5 Reactivity Studies with Mononuclear Complexes to Prepare Mixed Multimetallic Complexes

[((R,R)-C2Ph2L)(TiO)$_3$Ca(OTf)$_2$] ([14$_{Ca}$][(OTf)$_2$])

[14$_{Ca}$][(OTf)$_2$] was prepared similar to **[13$_{Ca}$][(OTf)$_2$]**. A mixture of TiO(acac)$_2$ (13.1 mg, 50.0 μmol, 3.10 equiv.) and **[9$_{Ca}$][(OTf)$_2$]** (22.0 mg, 16.1 μmol, 1.00 equiv.) in DMF (5 mL) was stirred for 19 h at r.t. The solution was filtered and collected. A red solid was precipitated through fast addition of Et_2O and n-pentane. The solid was collected and rinsed with Et_2O. The

Heterometallic Tetranuclear Complexes Featuring a Tris(ONNO)-Type Ligand

product **[14$_{Ca}$][(OTf)$_2$]** was obtained as a red solid. **FAB MS** (pos. mode, 3-NBA matrix): m/z = 1401.1 ([((R,R)-^{C2Ph2}L)(TiO)$_3$Ca(OTf)]$^+$), 1190.2 ([H((R,R)-^{C2Ph2}L)(TiO)$_2$Ca]$^+$), 1251.1 ([((R,R)-^{C2Ph2}L)(TiO)$_3$Ca]$^+$). **IR** in KBr: ($\tilde{\nu}$, cm^{-1}) 3442 (w), 3069 (w), 3029 (w), 2919 (w), 1638 (vs, C=N), 1597 (s), 1529 (vs), 1485 (s), 1466 (s), 1362 (s), 1273 (vs), 1242 (vs), 1221 (vs), 1156 (m), 1093 (w), 1030 (vs), 931 (w), 871 (w), 766 (m), 744 (m), 701 (s), 638 (s), 572 (m), 517 (m), 459 (m).

B.3.5 References

1. Clarke, R. M.; Storr, T. *Dalton Trans.* **2014**, *43*, 9380-9391.
2. Akine, S.; Utsuno, F.; Piao, S.; Orita, H.; Tsuzuki, S.; Nabeshima, T. *Inorg. Chem.* **2016**, *55*, 810-821.
3. Pokharel, U. R.; Fronczek, F. R.; Maverick, A. W. *Nat. Commun.* **2014**, *5*, 5883-5887.
4. Akine, S.; Nabeshima, T. *Dalton Trans.* **2009**, *47*, 10395-10408.
5. Tanifuji, K.; Ohki, Y., Recent Advances in the Chemical Synthesis of Nitrogenase Model Clusters. In *Structure and Bonding*, Springer: Berlin, Heidelberg, 2018.
6. Kaim, W.; Schwederski, B., Biologische Funktion der "frühen" Übergangsmetalle: Molybdän, Wolfram, Vanadium, Chrom. In *Bioanorganische Chemie*, 4 ed.; Vieweg+Teubner Verlag: Wiesbaden, 2005; pp 222-247.
7. Umena, Y.; Kawakami, K.; Shen, J.-R.; Kamiya, N. *Nature* **2011**, *473*, 55-61.
8. Kanady, J. S.; Tsui, E. Y.; Day, M. W.; Agapie, T. *Science* **2011**, *333*, 733-737.
9. Tsui, E. Y.; Agapie, T. *Proc. Natl. Acad. Sci. USA* **2013**, *110*, 10084-8.
10. Lee, H. B.; Shiau, A. A.; Oyala, P. H.; Marchiori, D. A.; Gul, S.; Chatterjee, R.; Yano, J.; Britt, R. D.; Agapie, T. *J. Am. Chem. Soc.* **2018**, *140*, 17175-17187.
11. Van Veggel, F. C. J. M.; Verboom, W.; Reinhoudt, D. N. *Chem. Rev.* **1994**, *94*, 279-299.
12. Verboom, W.; Rudkevich, D. M.; Reinhoudt, D. N. *Pure Appl. Chem.* **1994**, *66*, 679-686.
13. M. G. Antonisse, M.; N. Reinhoudt, D. *Chem. Commun.* **1998**, *0*, 443-448.
14. Van Veggel, F. C. J. M.; Harkema, S.; Bos, M.; Verboom, W.; Van Staveren, C. J.; Gerritsma, G. J.; Reinhoudt, D. N. *Inorg. Chem.* **1989**, *28*, 1133-1148.
15. Van Veggel, F. C. J. M.; Harkema, S.; Bos, M.; Verboom, W.; Woolthuis, G. K.; Reinhoudt, D. N. *J. Org. Chem.* **1989**, *54*, 2351-2359.
16. Van Veggel, F. C. J. M.; Bos, M.; Harkema, S.; Van de Bovenkamp, H.; Verboom, W.; Reedijk, J.; Reinhoudt, D. N. *J. Org. Chem.* **1991**, *56*, 225-235.
17. Van Veggel, F. C. J. M.; Bos, M.; Harkema, S.; Verboom, W.; Reinhoudt, D. N. *Angew. Chem. Int. Ed.* **1989**, *28*, 746-748.
18. Glaser, T.; Heidemeier, M.; Fröhlich, R.; Hildebrandt, P.; Bothe, E.; Bill, E. *Inorg. Chem.* **2005**, *44*, 5467-5482.
19. Feltham, H. L. C.; Clérac, R.; Ungur, L.; Vieru, V.; Chibotaru, L. F.; Powell, A. K.; Brooker, S. *Inorg. Chem.* **2012**, *51*, 10603-10612.
20. Feltham, H. L. C.; Clérac, R.; Powell, A. K.; Brooker, S. *Inorg. Chem.* **2011**, *50*, 4232-4234.
21. Feltham, H. L. C.; Clérac, R.; Ungur, L.; Chibotaru, L. F.; Powell, A. K.; Brooker, S. *Inorg. Chem.* **2013**, *52*, 3236-3240.
22. Feltham, H. L. C.; Clérac, R.; Peng, Y.; Moreno-Pineda, E.; Powell, A. K.; Brooker, S. *Z. Anorg. Allg. Chem.* **2018**, *644*, 775-779.
23. Feltham, H. L. C.; Dhers, S.; Rouzières, M.; Clérac, R.; Powell, A. K.; Brooker, S. *Inorg. Chem. Front.* **2015**, *2*, 982-990.
24. Feltham, H. L. C.; Klöwer, F.; Cameron, S. A.; Larsen, D. S.; Lan, Y.; Tropiano, M.; Faulkner, S.; Powell, A. K.; Brooker, S. *Dalton Trans.* **2011**, *40*, 11425-11432.
25. Kleemann, J. Homogenkatalysatoren für die Copolymerisation von Cyclohexenoxid mit CO_2 auf Basis multimetallischer Komplexe von Seltenerdmetallen und Zink. Doctoral Thesis, RWTH Aachen University, Aachen, 2017.
26. Nagae, H.; Aoki, R.; Akutagawa, S. N.; Kleemann, J.; Tagawa, R.; Schindler, T.; Choi, G.; Spaniol, T. P.; Tsurugi, H.; Okuda, J.; Mashima, K. *Angew. Chem. Int. Ed.* **2018**, *57*, 2492-2496.
27. Akine, S.; Sunaga, S.; Taniguchi, T.; Miyazaki, H.; Nabeshima, T. *Inorg. Chem.* **2007**, *46*, 2959-2961.
28. Nabeshima, T.; Miyazaki, H.; Iwasaki, A.; Akine, S.; Saiki, T.; Ikeda, C. *Tetrahedron* **2007**, *63*, 3328-3333.
29. Dhers, S.; Feltham, H. L. C.; Rouzières, M.; Clérac, R.; Brooker, S. *Dalton Trans.* **2016**, *45*, 18089-18093.

30. Feltham, H. L. C.; Lan, Y.; Klöwer, F.; Ungur, L.; Chibotaru, L. F.; Powell, A. K.; Brooker, S. *Chem. Eur. J.* **2011**, *17*, 4362-4365.
31. Shannon, R. D. *Acta Cryst. A* **1976**, *32*, 751-767.
32. Feltham, H. L.; Clerac, R.; Ungur, L.; Chibotaru, L. F.; Powell, A. K.; Brooker, S. *Inorg. Chem.* **2013**, *52*, 3236-3240.
33. Fulmer, G. R.; Miller, A. J. M.; Sherden, N. H.; Gottlieb, H. E.; Nudelman, A.; Stoltz, B. M.; Bercaw, J. E.; Goldberg, K. I. *Organometallics* **2010**, *29*, 2176-2179.
34. Bellemin-Laponnaz, S.; Dagorne, S., Coordination Chemistry and Applications of Salen, Salan and Salalen Metal Complexes. In *PATAI'S Chemistry of Functional Groups*, Rappoport, Z., Ed. John Wiley & Sons, Ltd.: 2012.
35. Rehder, D. *Metallomics* **2015**, *7*, 730-42.
36. Langeslay, R. R.; Kaphan, D. M.; Marshall, C. L.; Stair, P. C.; Sattelberger, A. P.; Delferro, M. *Chem. Rev.* **2018**, DOI: 10.1021/acs.chemrev.8b00245.
37. Pessoa, J. C. *J. Inorg. Biochem.* **2015**, *147*, 4-24.
38. Ando, R.; Ono, H.; Yagyu, T.; Maeda, M. *Inorg. Chim. Acta* **2004**, *357*, 817-823.
39. Fehrmann, R.; Boghosian, S.; Papatheodorou, G. N.; Nielsen, K.; Berg, R. W.; Bjerrum, N. J. *Inorg. Chem.* **1989**, *28*, 1847-1853.
40. Hagen, W. R. *Dalton Trans.* **2006**, *0*, 4415-4434.
41. Smith, T. S.; Lobrutto, R.; Pecoraro, V. L. *Coord. Chem. Rev.* **2002**, *228*, 1-18.
42. Schilder, H.; Lueken, H. *J. Magn. Magn. Mater.* **2004**, *281*, 17-26.
43. Speldrich, M.; Schilder, H.; Lueken, H.; Kögerler, P. *Isr. J. Chem.* **2011**, *51*, 215-227.
44. Son, S. B.; Skibida, I. P.; Maizus, Z. K. *Russ. Chem. Bull.* **1974**, *23*, 1890-1894.
45. Ulas, G.; Lemmin, T.; Wu, Y.; Gassner, G. T.; DeGrado, W. F. *Nat. Chem.* **2016**, *8*, 354-359.
46. Carney, J. R.; Dillon, B. R.; Thomas, S. P. *Eur. J. Org. Chem.* **2016**, *2016*, 3912-3929.
47. Spek, A. L. *J. Appl. Cryst.* **2003**, *36*, 7-13.
48. Gallant, A. J.; Yun, M.; Sauer, M.; Yeung, C. S.; MacLachlan, M. J. *Org. Lett.* **2005**, *7*, 4827-4830.
49. Akine, S.; Taniguchi, T.; Nabeshima, T. *Tetrahedron Lett.* **2001**, *42*, 8861-8864.
50. Liu, W.; Thorp, H. H. *Inorg. Chem.* **1993**, *32*, 4102-4105.
51. Brown, I. D.; Altermatt, D. *Acta Cryst. B* **1985**, *41*, 244-247.
52. Brese, N. E.; O'Keeffe, M. *Acta Cryst. B* **1991**, *47*, 192-197.
53. Atkins, P.; de Paula, J., *Physical Chemistry*. 9th ed.; Oxford University Press: Oxford, 2010.
54. Griffith, J. S., *The Theory of Transition-Metal Ions*. Cambridge University Press: Cambridge, 1980.
55. Gallant, A. J.; Hui, J. K. H.; Zahariev, F. E.; Wang, Y. A.; MacLachlan, M. J. *J. Org. Chem.* **2005**, *70*, 7936-7946.
56. MacLachlan, M. J. *Pure Appl. Chem.* **2006**, *78*, 873-888.
57. Grudzien, K.; Malinska, M.; Barbasiewicz, M. *Organometallics* **2012**, *31*, 3636-3646.

B.4 Catalytic Transformations of Heterocumulenes with Epoxides

B.4.1 Introduction

Catalytic conversion of heterocumulenes such as CO_2, CS_2 and isocyanates (RNCO) with epoxides leads to various compounds relevant to industrial application, such as aliphatic polycarbonates, cyclic organic carbonates (COC) and oxazolidinones (Scheme B.4.1).[1-3] Polycarbonates can be directly used in consumer products or as polyol blends in polyurethanes.[4] COCs can be applied in lithium-ion batteries[5, 6] as high-boiling-point solvents[7] with high dielectric constants. Oxazolidinones are of interest as they are components of pharmaceuticals[8] and synthetic intermediates[9]. Suitable catalysts are required to selectively produce the desired reaction products.

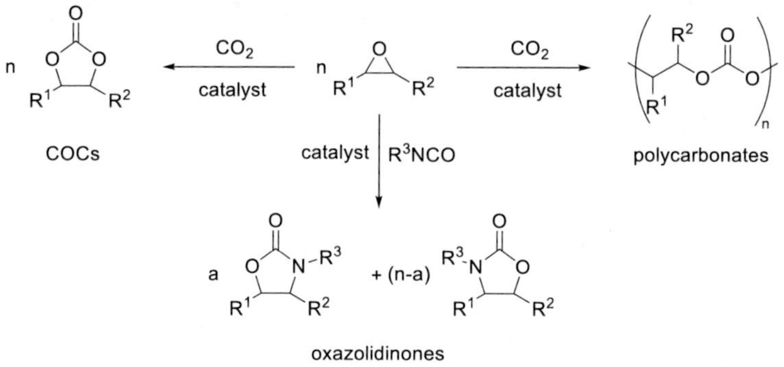

Scheme B.4.1. Catalytic conversion of heterocumulenes with epoxides to produce COCs, polycarbonates and oxazolidinones.

The reaction of heterocumulenes with epoxides is currently dominated by the synthesis of COCs and polycarbonates and has received increased attention since the year 2000.[2, 3] Aliphatic polycarbonates can be produced through ring-opening co-polymerization of epoxides with CO_2.[3] These polymers are often biodegradable and may potentially replace conventional polymers based on phosgene.[10] Since the pioneering work by Inoue on the co-polymerization of propylene oxide (PO) with CO_2 catalyzed by $ZnEt_2/H_2O$ in the late 1960s, numerous co-polymerization catalysts haven been developed.[11] Most research has focused on homogeneous catalyst systems as heterogeneous ones often require high temperatures and elevated pressures of CO_2, rendering them often unfeasible for industrial application.[12] The group of homogeneous catalysts can be divided into two classes: The first class consists of mononuclear catalysts (Figure B.4.1) and the second one of multinuclear catalysts which rely on metal cooperativity (Figure B.4.2).[3, 12, 13] A large number of catalysts have been designed featuring a variety of different metal centers. In case mononuclear catalysts are used, a binary system is employed composed of the metal catalyst (Figure B.4.1(a) and (b)) and a co-catalyst which provides the nucleophile.[3, 12, 14-18] Common co-catalysts are Lewis bases (e.g. DMAP) or organic salts (e.g. tetrabutylammonium salts) which act as phase transfer agents and provide nucleophiles for the ring-opening step.[3, 12] The co-catalyst may be dropped into the reaction mixture or may be intramolecularly provided as an anchored functional group of the ligand backbone (Figure B.4.1(c)).[19]

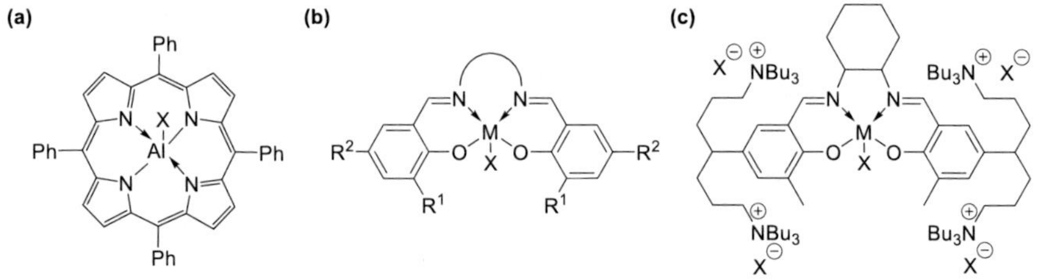

Figure B.4.1. Selected mononuclear catalysts featuring (a) porphyrin-type ligands,[14-17] (b) salen-type ligands[18] and (c) bifunctional[19] salen-type ligands.[12]

Contrary to mononuclear catalysts, multinuclear catalysts commonly feature all relevant components required for the catalytic co-polymerization within one molecule, namely a Lewis acid to activate the epoxide and a nucleophile to ring-open the incoming epoxide (Figure B.4.2).[3, 12, 20] Due to cooperative effects of the metal centers, these complexes commonly exhibit a higher catalytic activity than mononuclear binary systems. Williams and co-workers developed several heterodinuclear catalysts and have clarified the mechanism of dinuclear catalysts in co-polymerization reactions (Figure B.4.2(a)).[12] Many of these dinuclear catalysts contain zinc.

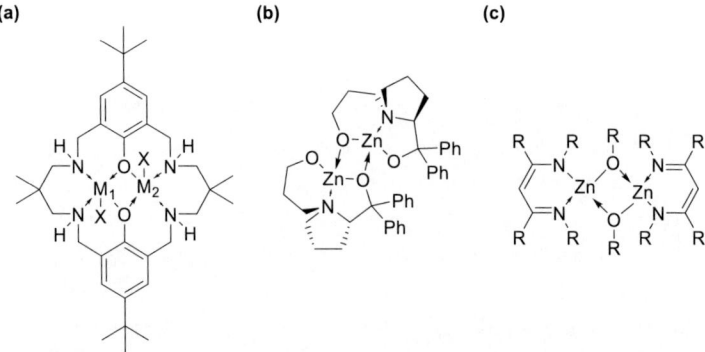

Figure B.4.2. Selected dinuclear catalysts stabilized by (a) Robson-type ligands,[12] (b) proline ligands[21] and (c) β-diketiminate ligands[12].

Results and Discussion

Mechanistic investigations by Williams and co-workers on dinuclear catalysts revealed a chain shuffling mechanism in which the growing polymer chain is alternately coordinated at one of the two catalytic centers M_1 and M_2 (Scheme B.4.2).[22, 23] The incoming cyclohexene oxide (CHO) is coordinated at M_2 and activated. Ring-opening of the epoxide by a provided nucleophile produces an alkoxo species. Subsequent insertion of CO_2 generates a carbonate species at the second metal center M_1 which is bridged between the two metal centers due to dative interaction between the carbonyl oxygen and M_2. Coordination of an additional epoxide molecule occurs at M_2, transferring the polymer chain to M_1. Chain propagation eventually produces poly(cyclohexene carbonate) (PCHC). Backbiting or depolymerization occur as side reactions, producing COCs.

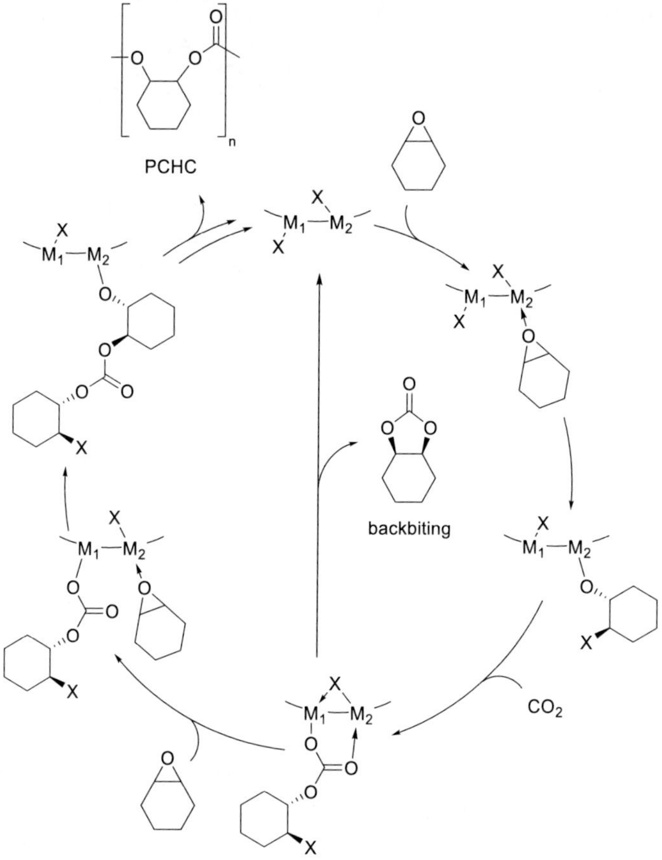

Scheme B.4.2. Catalytic cycle proposed based on dinuclear catalysts by Williams and co-workers, rationalizing formation of cis-CHC and PCHC.[22]

Detailed mechanistic and kinetic investigations indicated that PCHC is the kinetic product whereas CHC is the thermodynamic product.[23] Commonly, CHO is used for mechanistic studies on the co-polymerization reaction due to its decreased tendency to produce cyclohexene carbonate (CHC). CHC formation is favored over PCHC formation in the presence of good leaving groups and at elevated temperatures and pressures. [22, 23]

Since COCs are value-added synthetic intermediates, various catalytic systems have been developed to selectively produce these compounds.[1, 2] Many of these catalysts are similar to reported co-polymerization catalysts and a large library of heterogeneous and homogeneous catalysts exists;[2] the latter class is based on organic molecules,[6, 24] organic halide salts,[25] main group metal complexes[26, 27] and transition metal[28] complexes. Recently, simple catalysts based on amines, Schiff bases[29] and *in situ* generated main group metal 18c6 complexes[26, 30] (Figure B.4.3(a)) emerged. Drawbacks of these systems are often higher catalyst loadings or lower *cis/trans* selectivities compared to more sophisticated catalysts. Mashima and co-workers also reported a highly active tetranuclear zinc cluster which produces COCs under mild reaction conditions, which requires a co-catalyst (Figure B.4.3(b)).[31] Vanadium based-catalysts have also been applied in catalytic COC formation from epoxides and CO_2, albeit with moderate activity (Figure B.4.3(c)).[2, 22, 32, 33] Similar to the co-polymerization reaction, these catalysts generally require a nucleophile which ring opens the epoxide, and a good leaving group to facilitate the cyclization reaction.[2]

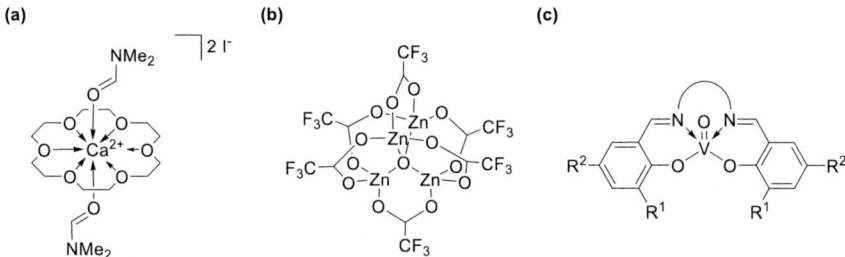

Figure B.4.3. Selected metal-based catalysts for the formation of COCs from epoxides and CO_2 based on (a) a calcium 18c6 complex, [26, 30] (b) a tetranuclear zinc cluster[31] and (c) a mononuclear vanadyl complex[32, 33] featuring a salen-type ligand.[2]

COCs and oxazolidinones are structurally related: Exchanging one oxygen atom of the five-membered ring heterocycle against an amine affords an oxazolidinone framework. Oxazolidinones can be prepared from epoxides and isocyanates or for certain substrates directly from amines, epoxides and CO_2.[34-36] Catalysts based on organometallic reagents, ammonium salts or metal salts have been employed in the catalytic formation of oxazolidinones from epoxides and isocyanate.[37-42] Recently, aluminum and vanadium complexes featuring

Results and Discussion

salen-type ligands have also been applied in this reaction.[43] Regio- and enantioselectivity remain key issues as most catalysts produce a mixture of the two regiomers which often also exist as racemates.

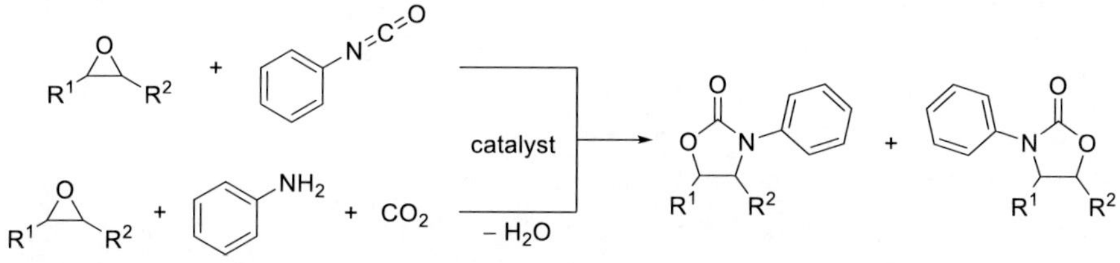

Scheme B.4.3. Catalytic conversion of epoxides with phenyl isocyanate or aniline and CO_2 to produce oxazolidinones. [34-36]

Various mononuclear catalysts featuring salen-type ligands haven been employed in the catalytic conversion of heterocumulenes with epoxides.[1, 12, 18] To enhance their catalytic performance through metal cooperativity, catalysts featuring macrocyclic tris(ONNO)-type ligands have been developed. Recently, Mashima, Okuda and co-workers applied tetranuclear lanthanide trizinc complexes stabilized by tris(ONNO)-type ligands in the co-polymerization of CHO with CO_2 to produce polycarbonates with high carbonate linkages.[44, 45]

B.4.2 Results and Discussion

Inspired by the recent results of Mashima, Okuda and co-workers on the application of tetranuclear macrocyclic complexes in the co-polymerization of CHO with CO_2, complexes synthesized in sections B.2 and B.3 were tested for their catalytic activity regarding the reaction of heterocumulenes (CO_2 and phenyl isocyanate) with epoxides.[44, 45] Initial testing of the catalysts was performed under neat conditions with CHO, CO_2 (p_{CO2} = 10 bar) and the respective catalyst ($n_{catalyst}/n_{epoxide}$ = 2000) at 100 °C. All experiments were conducted in nickel alloy autoclaves. Samples of the reaction mixture were analyzed by ^{1}H NMR spectroscopy and size exclusion chromatography (SEC) according to literature procedures.[44-48] The anticipated products of the reaction are the co-polymerization product poly(cyclohexene carbonate) (PCHC), the ring-opening polymerization product polyether and the insertion product cyclohexene carbonate (CHC). The methine protons of the polycarbonate units are observed at δ 4.65 ppm and the ones of the polyether units at δ 3.45 ppm in the ^{1}H NMR spectrum in chloroform-d_1. The selectivity of PCHC compared to polyether is given as the carbonate linkage which describes the ratio of polycarbonate vs. polyether bonds as determined by the integral values of the respective signals in the ^{1}H NMR spectra. Cyclic cyclohexene carbonate can be obtained as a cis- or trans-product with resonances for the methine protons at δ 4.68 and 4.00 ppm, respectively.[22]

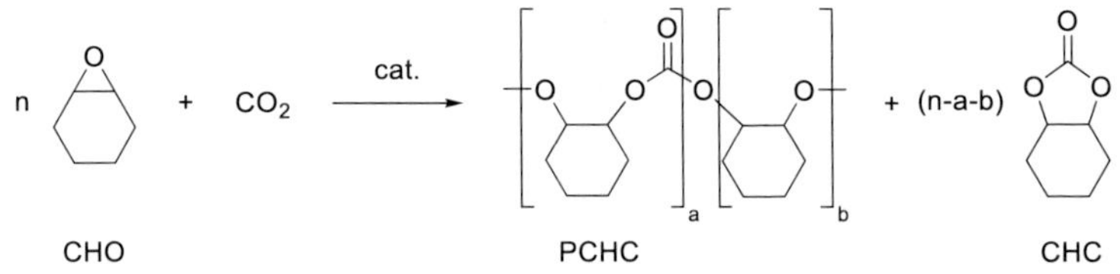

Scheme B.4.4. Reaction of CHO with CO_2 to produce poly(cyclohexene carbonate) (PCHC), polyether and cyclohexene carbonate (CHC).

Table B.4.1 shows the results of the initial catalyst screening. The mononuclear complexes [9$_{Ca}$][(OTf)$_2$], [9$_{Ca}$][I$_2$] and [9$_{La}$][(OTf)$_3$] did not exhibit any catalytic activity (Entries 1 – 3). When the reaction was performed in MeCN and at a higher catalyst loading of 5 mol%, again no catalytic activity was observed and only the starting material was detected by NMR spectroscopy. The negligible catalytic activity of the mononuclear complexes can be rationalized by the low solubility of the catalysts and by the absence of an additional Lewis acid. The Lewis acid is required for binding CO_2 and facilitating formation of the co-polymerization products. However, these results are quite surprising, since Werner and co-workers have previously reported a mononuclear calcium iodide complexes featuring an 18c6 ligand which converts epoxides to cyclic carbonates at high yields under mild reactions conditions without a co-catalyst.[26]

Table B.4.1. Results of the initial catalyst screening for the catalytic conversion of CHO with CO_2.

entry[a]	catalyst	metals A_3B	conversion[b] [%]	polymer[b] [%]	carbonate linkage[b] [%]	CHC[b] [%]
1	[9_{Ca}][(OTf)$_2$]	Ca	<1	0	/	<1
2[c]	[9_{La}][(OTf)$_3$]	La	<1	0	/	<1
3[c]	[9_{Ca}][I$_2$]	Ca	<1	0	/	<1
4[d]	[11_{Ca}][I$_2$]	Zn_3Ca	>99	>99	80	<1
5	[13_{Ca}][I$_2$]	$(VO)_3Ca$	28	<1	0	>98 cis <1 trans
6[e]	[16_{Ca}][Cl$_3$]	Mn_3Ca	>99	>99	50	<1

[a] cat. (0.05 mol%), CHO (20 mmol), CO_2 (10 bar), 100 °C, 24 h; [b] determined by ^{1}H NMR spectroscopy in chloroform-d_1; [c] cat. (5 mol%), in MeCN (2 mL); [d] cat. (0.1 mol%); [e] cat. (0.1 mol%), CHO (10 mmol), CO_2 (15 bar), 120 °C.

The heterometallic tetranuclear complexes [11_{Ca}][I$_2$], [13_{Ca}][I$_2$] and [16_{Ca}][Cl$_3$] exhibited catalytic activity at the tested reaction conditions, although different selectivities were observed (Table B.4.1, Entries 4 – 6). At a catalyst loading of 0.1 mol%, [11_{Ca}][I$_2$] completely polymerized CHO with a carbonate linkage of 80% and a polyether content of 20% (Entry 4). The back-biting product CHC was not observed, indicating that depolymerization did not occur. Similar to [11_{Ca}][I$_2$], only the polymerization product was observed when [16_{Ca}][Cl$_3$] was used as a catalyst (Entry 6). At a catalyst loading of 0.1 mol% and at a CO_2 pressure of 15 bar, [16_{Ca}][Cl$_3$] completely converted CHO and exclusively produced the polymerization product with a carbonate linkage of 50%. Due to performing the reaction under neat conditions until completion of the polymerization reaction, the viscosity of the reaction mixture is so high that agitation and CO_2 diffusion are inhibited, resulting in a lower carbonate linkage. Surprisingly, [13_{Ca}][I$_2$] produces cis-CHC with high selectivity at a conversion of 28% (Entry 5). Analysis of the reaction mixture with SEC revealed only trace amounts of oligomeric species.

Previously, tetranuclear trizinc complexes featuring a [3+3] macrocyclic ligand were used to selectively obtain the co-polymerization product with high carbonate linkage.[44, 45] Exchanging the rare earth metals against the more abundant calcium cation, reduces the catalyst cost and may make the catalysts more relevant for industrial applications. To compare previous results to the ones obtained with the trizinc catalysts, prepared in section B.3.2.1, the co-polymerization experiments were repeated under similar conditions. Since trizinc catalysts featuring acetate anions were previously reported to exhibit high catalytic activity, [11_{Ca}][(OAc)$_2$] was also tested as catalyst for the co-polymerization reaction. The reactions were performed at a catalyst loading of 0.05 mol% and stopped after 3 h. Table B.4.2 summarizes the obtained results of the co-polymerization reaction. An important parameter to evaluate the catalytic performance is the turnover frequency (TOF) which describes the amount of CHO consumed per mole of catalyst and per time. Both catalysts selectively produce PCHC with a high carbonate linkage of >99%. [11$_{Ca}$][I$_2$] shows a higher catalytic activity with a TOF of 10 h^{-1} (Entry 1) compared to [11$_{Ca}$][(OAc)$_2$] with a TOF of 7 h^{-1} (Entry 2). The PCHC product obtained with [11$_{Ca}$][I$_2$] has a slightly higher molecular weight compared to the one obtained with [11$_{Ca}$][(OAc)$_2$], although both have a similar narrow molecular weight distribution with a polydispersity index (PDI) of 1.1 – 1.2. Since resonances for syndiotactic and isotactic units were observed at δ 153.2 and 153.7 ppm in the respective ^{13}C{^{1}H} NMR spectra in chloroform-d_1, the obtained polymers are atactic. These results are similar to the ones obtained for the tetranuclear lanthanum trizinc complex ([5$_{La}$][(OAc)$_3$] reported in literature, although ([5$_{La}$][(OAc)$_3$] exhibits a higher activity with a TOF of 333 h^{-1} (Entry 3).[45] As the catalytic activity of [11$_{Ca}$][(OAc)$_2$] and [11$_{Ca}$][I$_2$] is lower than the of ([5$_{La}$][(OAc)$_3$] further research on these catalysts was discontinued.

Table B.4.2. Results of the catalytic co-polymerization of CHO with CO_2.

entry[a]	catalyst	conversion[b] [%]	carbonate linkage[b] [%]	M_n (PDI)[c] [10^4 g mol^{-1}]	TOF[b] [h^{-1}]
1	[11$_{Ca}$][I$_2$]	12	>99	1.4 (1.2)	10
2	[11$_{Ca}$][(OAc)$_2$]	8	>99	1.2 (1.1)	7
3[45]	[11$_{La}$][(OAc)$_3$]	50	99	1.0 (1.2)	333

[a] cat. (0.05 mol% in respect to CHO), CHO (20 mmol), CO_2 (10 bar), 100 °C, 3 h; [b] determined by ^{1}H NMR spectroscopy in chloroform-d_1; [c] determined by SEC in THF.

Only few vanadium catalysts for the formation of COCs have been reported in literature. To understand the reason for the preferential formation of the CHC product over the co-polymerization product when exchanging zinc against vanadium, further experiments were performed using trivanadium complexes prepared in section B.3.2.2 as catalysts.[32] The reactions were performed at a higher catalyst loading of 0.1 mol% to increase the conversion from 28% to 41% in the case of [13$_{Ca}$][I$_2$] as a catalyst (Table B.4.3, entry 1). The selectivity regarding the formation of cis-CHC remained unchanged. To investigate the influence of the anion onto the reaction, calcium trivanadyl complexes [((R,R)-^{C2Ph2}L)(VO)$_3$CaX$_2$] with X = Br ([13$_{Ca}$][Br$_2$]), Cl ([13$_{Ca}$][Cl$_2$]) and OTf ([13$_{Ca}$][(OTf)$_2$]) were used as catalysts (Table B.4.3, Entries 2 – 4). In case of [13$_{Ca}$][(OTf)$_2$], only the undesired polyether product is generated and CHC or PCHC are not detected (Entry 4). Blank experiments were performed without adding any catalyst or with addition of the starting materials CaI$_2$, VO(acac)$_2$ or a combination of both (Table B.4.3, Entries 5 – 8). In all cases only a low activity with a TOF value of <1 h^{-1} was observed. In case of CaI$_2$, a low amount of cis-CHC was produced with poor CHC/ether selectivity. In case of VO(acac)$_2$ and a combination of CaI$_2$ with VO(acac)$_2$, only the ether side products were observed. In case no additional catalyst was added to the reaction, no catalytic activity was observed. The blank experiments exclude any significant contribution of impurities, starting materials or the surface of the autoclave onto the observed catalytic activity.

Table B.4.3. Results of the anion screening in the catalytic formation of CHC from CHO and CO$_2$ and blank experiments.

entry[a]	catalyst	conversion[b] [%]	CHC[c] [%]	ether bonds[c] [%]	TOF[b] [h^{-1}]
1	[13$_{Ca}$][I$_2$]	41	>98 cis <1 trans	1	17
2	[13$_{Ca}$][Br$_2$]	26	89 cis 7 trans	4	11
3	[13$_{Ca}$][Cl$_2$]	5	51 cis 22 trans	27	2
4	[13$_{Ca}$][(OTf)$_2$]	29	0	>99	12
5[d]	CaI$_2$	3	64 cis 0 trans	36	<1
6[e]	VO(acac)$_2$	<1	0	>99	<1
7[f]	CaI$_2$ + 3 VO(acac)$_2$	<1	0	>99	<1
8	/	0	0	0	/

[a] cat. (0.1 mol% in respect to CHO), CHO (10 mmol), CO$_2$ (10 bar), 100 °C, 24 h; [b] determined by the integral ratios of consumed CHO and the sum of the integrals of the products and CHO in the ^{1}H NMR spectra in chloroform-d_1; [c] selectivity determined by the integral ratios of the respective products and the sum of the products in the ^{1}H NMR spectra in chloroform-d_1; [d] cat. (0.05 mol%); [e] cat. (0.15 mol%), CHO (20 mmol); [f] CaI$_2$ + 3 VO(acac)$_2$ (0.05 mol%), CHO (20 mmol).

When exchanging iodide against the smaller halogen anions bromide and chloride, the *cis-/trans*-selectivity and catalytic activity (TOF) decrease in the order I⁻ > Br⁻ > Cl⁻. (Table B.4.3, Entries 1 – 3) Concomitantly, more undesired side reactions occur, and the amount of ether bonds increases in the same order. The change of selectivity and activity is rationalized by a decreasing nucleophilicity when going from iodide to chloride. Plotting the TOF values against the pK_B values as an index of the nucleophilicity of the halogen anions, a linear trend is observed (Figure B.4.4).[49] In case of the halogen anions, the catalytic activity linearly increases with increasing nucleophilicity. A dependence of the catalytic activity on the nature of the nucleophilic anion is commonly observed in catalytic transformations of epoxides with CO_2.[12, 22, 45] Since **[13$_{Ca}$][I$_2$]** exhibited the highest selectivity and activity, iodide was selected as a counter anion for further investigations.

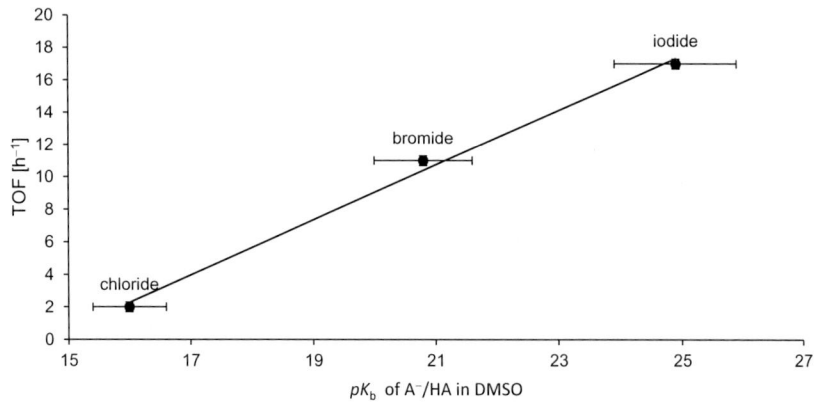

Figure B.4.4. Graphical representation of the catalytic activity (TOF) vs. the pK_B values of the anions and their corresponding acids in DMSO (R^2 = 0.9952).[49]

To increase the catalytic activity and selectivity, the temperature and CO_2 pressure were optimized with a 2^2 full factorial experimental design with one center point. Optimization of the temperature and CO_2 pressure was chosen, as the influence of these variables onto the selectivity of CHC over PCHC has been previously reported.[23] 80 °C (–) and 5 bar (–) were selected as bottom points and 120 °C (+) and 15 bar (+) as top points on the experimental grid. The starting conditions 100 °C and 10 bar CO_2 pressure were selected as center point (o). The results of the optimization experiments are summarized in Table B.4.4. Performing the reaction at 80 °C (Entries 1 – 2), lower activity and selectivity were observed with little influence of the pressure at 5 (– –) or 15 bar (– +). An increase of the temperature to 120 °C results in a higher

Results and Discussion

catalytic activity (Entries 4 – 5). At a lower pressure of 5 bar (+ –), the catalytic activity is comparable to the center point (o). At an elevated pressure of 15 bar (+ +), the catalytic activity increases to a TOF of 34 h^{-1} while the selectivity remains unchanged. Similar behavior has been previously reported for several multinuclear catalytic systems: Here, the initial rates of CHC formation increase with increasing temperature due to a higher activation barrier for the thermodynamic CHC product compared to the kinetic PCHC product.[23, 50] The catalytic activity at optimized reaction conditions is higher compared to other vanadium catalysts.[33, 51] The mononuclear vanadium(V) complex featuring an amino-triphenolate ligand reported by Kleij, Licini and co-workers exhibits a TOF of approximately 11 h^{-1} with [^{n}Bu$_4$N][Br] as co-catalyst (85 °C, 10 bar, 0.5 mol%, 18 h, CHO (4 mmol)).[51] Further experiments with the trivanadyl catalytic system were performed under optimized reaction conditions.

Table B.4.4. Optimization of reaction conditions to selectively produce cis-CHC at high conversions of CHO.

CHO + CO$_2$ —[13$_{Ca}$][I$_2$]→ [polymer]$_m$ + b (cis-CHC) + (n-m-b) (trans-CHC)

entry[a]	factorial	T [°C]	p$_{CO2}$ [bar]	conversion[b] [%]	CHC[c] [%]	ether bonds[c] [%]	TOF[b] [h^{-1}]
1	– –	80	5	29	95 cis 1 trans	4	12
2	– +	80	15	25	95 cis 0 trans	5	10
3	o	100	10	41	>98 cis <1 trans	1	17
4	+ –	120	5	43	96 cis 2 trans	2	18
5	+ +	120	15	83	>98 cis <1 trans	<1	34

[a] [13$_{Ca}$][I$_2$] (0.1 mol% in respect to CHO), CHO (10 mmol), 24 h; [b] determined by the integral ratios of consumed CHO and the sum of the integrals of the products and CHO in the ^{1}H NMR spectra in chloroform-d_1; [c] selectivity determined by the integral ratios of the respective products and the sum of the products in the ^{1}H NMR spectra in chloroform-d_1.

To determine the influence of the central cation onto the catalytic reaction, the tetranuclear complexes [((R,R)-C2Ph2L)(VO)$_3$MI$_2$] (M = Sr for [13$_{Sr}$][I$_2$] , M = Ba for [13$_{Ba}$][I$_2$]) were prepared analogously to [13$_{Ca}$][I$_2$] from the corresponding mononuclear complexes [9$_{Sr}$][I$_2$] and [9$_{Ba}$][I$_2$]. [13$_{Sr}$][I$_2$] and [13$_{Ba}$][I$_2$] are paramagnetic and do not produce any resonances in the ^{1}H NMR spectra. The IR spectra of the compounds confirm the formation of the trivanadyl complexes as indicated by absorptions in the range of $\tilde{v}$ = 1609 – 1613 cm^{-1} for the imine bonds and at $\tilde{v}$ = 989 – 1002 cm^{-1} for the V=O bonds. The FAB mass spectra show peaks at m/z = 1308.5 ([13$_{Sr}$][I$_2$]) and 1358.5 ([13$_{Ba}$][I$_2$]) which correspond to the [M – 2I]$^+$ fragments, indicating successful formation of the complexes.

[13$_{Sr}$][I$_2$] and [13$_{Ba}$][I$_2$] were tested at the optimized reaction conditions and compared to [13$_{Ca}$][I$_2$] (Table B.4.5). The tetranuclear strontium complex [13$_{Sr}$][I$_2$] (Entry 2) exhibits similar catalytic activity and selectivity regarding the formation of cis-CHC as [13$_{Ca}$][I$_2$] with a TOF of 34 h^{-1} (Entry 1). The analogous barium complex [13$_{Ba}$][I$_2$] (Entry 3) shows only 2/3 of the activity of [13$_{Sr}$][I$_2$] and [13$_{Ca}$][I$_2$], albeit with similar cis/trans selectivity. The drop of catalytic activity when changing the central cation from calcium to barium is caused by a stronger interaction between barium and iodide as expected from the HSAB principle. Since there is little difference in the catalytic performance of [13$_{Ca}$][I$_2$] and [13$_{Sr}$][I$_2$], the tetranuclear calcium complex [13$_{Ca}$][I$_2$] was selected as the best catalyst.

Table B.4.5. Results of the cation screening on the formation of CHC from CHO and CO$_2$.

entry[a]	catalyst	conversion[b] [%]	CHC[c] [%]	ether bonds[c] [%]	TOF[b] [h^{-1}]
1	[13$_{Ca}$][I$_2$]	83	>98 cis <1 trans	<1	34
2	[13$_{Sr}$][I$_2$]	81	>99 cis 0 trans	<1	34
3	[13$_{Ba}$][I$_2$]	54	>98 cis <1 trans	<1	23

[a] cat. (0.1 mol% in respect to CHO), CHO (10 mmol), CO$_2$ (15 bar), 120 °C, 24 h; [b] determined by the integral ratios of consumed CHO and the sum of the integrals of the products and CHO in the ^{1}H NMR spectra in chloroform-d_1; [c] selectivity determined by the integral ratios of the respective products and the sum of the products in the ^{1}H NMR spectra in chloroform-d_1.

Results and Discussion

The catalyst [13$_{Ca}$][I$_2$] was used for the formation of cyclic organic carbonates (COCs) **H–P** from CO$_2$ and terminal epoxides **A – I** and internal epoxides **J – R** (Figure B.4.5). The conversions and selectivities were determined by ^{1}H NMR spectroscopy of the reaction mixtures according to literature procedures.[51, 52]

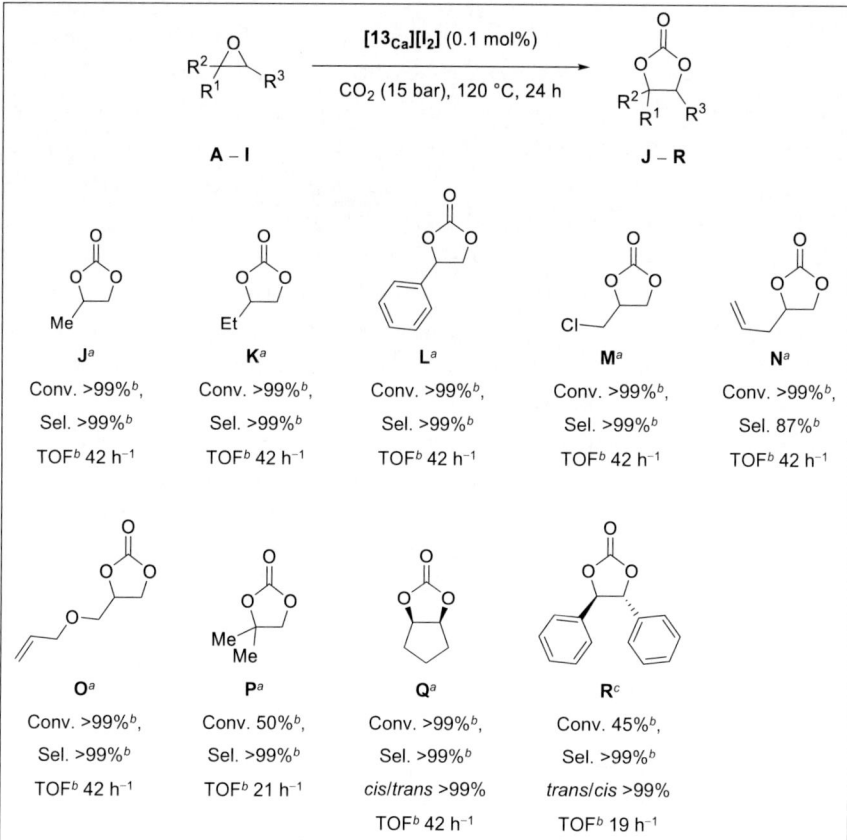

Figure B.4.5. Synthesis of COCs **H–P** from epoxides **A–G** and CO$_2$ using the catalyst [13Ca][I2]; [a] [13Ca][I2] (0.1 mol%), CHO (10 mmol), CO$_2$ (15 bar), 120 °C, 24 h; [b] determined by ^{1}H NMR spectroscopy in chloroform-d$_1$ according to literature procedures;[51, 52] [d] in MeCN (5 mL).

The conversion of the terminal epoxide propylene oxide (PO) **A** was first evaluated: [13$_{Ca}$][I$_2$] provided complete conversion of **A** to the COC product **J**. Full conversion of PO to propylene carbonate is also observed at milder reaction conditions after performing the reaction for 24 h at 80 °C with a pressure of 5 bar. Motivated by these results, the other epoxides featuring various functional groups were tested. [13$_{Ca}$][I$_2$] typically converts the epoxides quantitatively and selectively to their corresponding COCs. When the epoxide contains unsaturated side chains as in **E**, side product formation occurs when **E** is converted to **N** due to presumable olefin oligomerization reactions. When the terminal olefin is farther away from the epoxide reaction center as in **F**, less side reactions occur, and **O** is obtained with high selectivity. When the bulkiness around the epoxide is increased due to sterically demanding substituents as in **G** and **I**, the catalytic activity drops to a TOF of 21 and 19 h^{-1}, respectively. Since trans-stilbene oxide **I** is a solid at room temperature, the viscosity is expected to be high under neat conditions: Wen **I** is converted to **R** under neat conditions, only 8% of the starting material is converted. When the same reaction is performed in solution, the activity increases from 3 h^{-1} to 19 h^{-1}. A similar trend was reported for other systems as well.[51] The epoxide **H** with a smaller ring size compared to CHO, could also be quantitatively and selectively converted to the corresponding COC **Q**. For the internal epoxides **H** and **I**, different cis/trans selectivities are observed. **Q** is obtained similar to CHC exclusively as the cis-product, whereas **R** is exclusively obtained as the trans-product due to the sterically demanding phenyl substituents.[26, 51] The catalytic activity of **[13$_{Ca}$][I$_2$]** is higher than the one of the mononuclear calcium iodide complex reported by Werner and co-workers with a TOF of 0.4 h^{-1} in the conversion of CHO to CHC.[26] Under optimized reaction conditions, the activity of **[13$_{Ca}$][I$_2$]** also surpasses the one of the mononuclear vanadium catalyst reported by Kleij and Licini and co-workers.[51] The results highlight the activity of the calcium trivanadyl complex **[13$_{Ca}$][I$_2$]** in catalytic transformations of epoxides to COCs with a wide substrate scope and without the need of an additional co-catalyst.

A chain-shuffling mechanism similar to other multinuclear systems is proposed for the selective formation of cis-CHC from CHO catalyzed by **[13$_{Ca}$][I$_2$]** (Scheme B.4.5).[22, 45] In the first step (**i**) CHO coordinates to Ca^{2+} in the apical position, similar to the apical ancillary solvent ligands in the crystal structures of **[9$_{Ca}$][(OTf)$_2$]** and **[11$_{Ca}$][(OAc)$_2$]** (refer to sections B.2.2, B.3.2.1). In a subsequent step, iodide presumably bound to a vanadium center, (**ii**) ring-opens the epoxide to generate a Ca-alkoxide species.[45] The coordination of iodide to vanadium is anticipated by EPR studies, which indicated the formation of inequivalent vanadium centers in case of **[13$_{Ca}$][I$_2$]** (section B.3.2.2). In the insertion step, (**iii**) CO_2 coordinates to vanadium and (**iv**) the alkoxide reacts with CO_2 resulting in the formation of a carbonate species bound to vanadium. Since CHC formation was not observed when using the mononuclear complex **[9$_{Ca}$][I$_2$]**, it is expected that vanadium is required as a Lewis acid to stabilize CO_2. In the last step, backbiting occurs to produce cis-CHC, regenerating the catalyst. The backbiting mechanism is favored by more ionic metal-alkoxide bonds and the presence of good leaving groups such as iodide.[22] Whereas a depolymerization mechanism of PCHC would mostly produce trans-CHC.

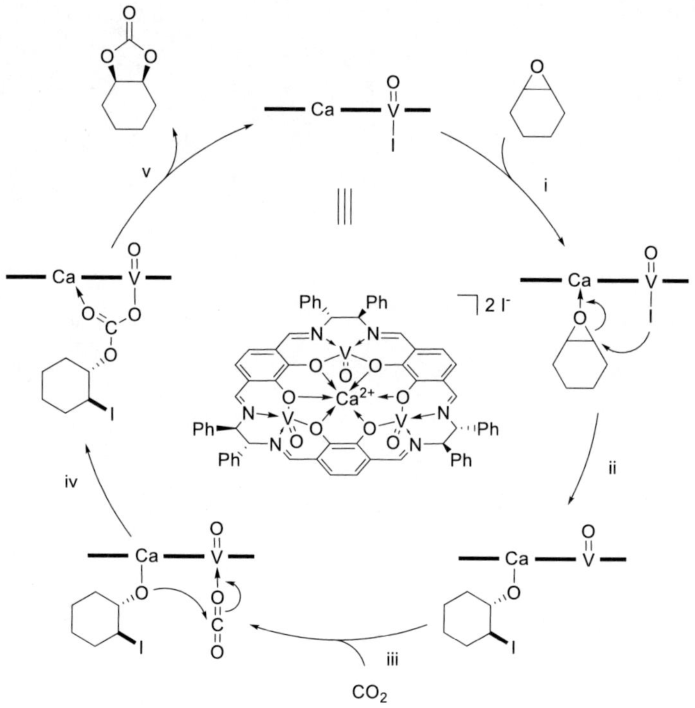

Scheme B.4.5. Proposed catalytic cycle rationalizing the formation of cis-CHC from CHO.

The reaction of epoxides with heterocumulenes may give access to a broad range of five-membered heterocycles such as the here reported COCs and oxazolidinones.[43] Motivated by the results on the catalytic reaction between epoxides and CO_2 to produce COCs, phenyl isocyanate was tested as an alternative heterocumulene substrate. Initial testing of [13$_{Ca}$][I$_2$] was performed under neat reaction conditions with stochiometric amounts of phenyl isocyanate and PO at a catalyst loading of 0.1 mol%. Samples of the reaction mixtures were analyzed by ^{1}H NMR spectroscopy and assigned according to literature.[43] Upon addition of the isocyanate to the solution of [13$_{Ca}$][I$_2$] in PO, the reaction mixture solidified within 10 min at room temperature due to fast reaction between PO and isocyanate to produce the oxazolidinone (melting point of 78 – 80 °C).[43] After a reaction time of an additional 30 min at 80 °C, 64% of the initial PO was regioselectively converted to the product 3-phenyl-5-methyloxazolidin-2-one, **Sa** (Scheme B.4.6). The methyl substituent of the 3,5-product **Sa** is observed at δ 1.52 ppm in the ^{1}H NMR spectrum in chloroform-d_1, whereas the 3,4-regioisomer **Sb** would exhibit a signal at δ 1.25 ppm which was not detected. In addition to the high regioselectivity, the catalyst exhibits a high activity with a TOF of 1280 h^{-1}. In case of a

Catalytic Transformations of Heterocumulenes with Epoxides

mononuclear vanadium(V) salen complex with [^{n}Bu$_4$N][Br] as co-catalyst, only a regioselectivity of 80% with a TOF of 8 h^{-1} is reported.[43]

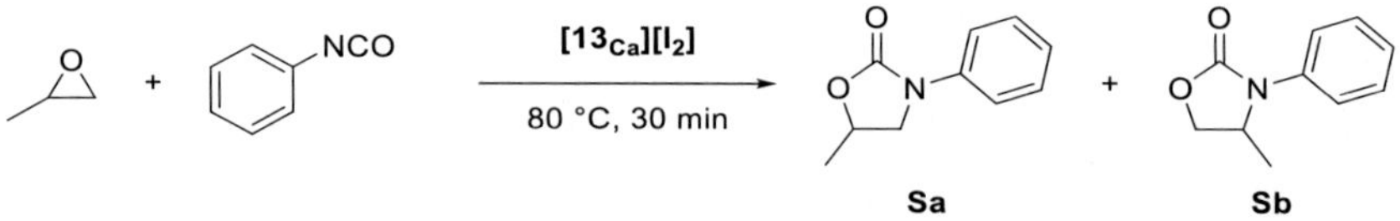

Scheme B.4.6. Regioselective formation of the oxazolidinone **Sa** from PO and phenyl isocyanate.

B.4.3 Summary and Outlook

Mono- and tetranuclear complexes synthesized in sections B.2 and B.3 were tested for their catalytic activity regarding the reaction of heterocumulenes (CO_2 and phenyl isocyanate) with epoxides. The mononuclear complexes [9$_{Ca}$][(OTf)$_2$], [9$_{Ca}$][I$_2$] and [9$_{La}$][(OTf)$_3$] are not catalytically active in the reaction of CHO with CO_2. When additional transition metals are coordinated to the complexes, the tetranuclear complexes become catalytically active but exhibit different selectivities depending on the used metal combinations: The trimanganese and trizinc catalysts favor formation of the polymeric products, whereas the trivanadyl complexes favor conversion of epoxides to COCs.

The calcium trizinc complexes [11$_{Ca}$][(OAc)$_2$] and [11$_{Ca}$][I$_2$] selectively co-polymerize CHO with CO_2 to poly(cyclohexene carbonate) with high carbonate linkage. The properties of the resulting polymers are similar to the ones obtained with ([5$_{La}$][(OAc)$_3$].[44, 45] However, the catalytic activity of [11$_{Ca}$][(OAc)$_2$] and [11$_{Ca}$][I$_2$] is only 3% of the one exhibited by ([5$_{La}$][(OAc)$_3$].

The trivanadyl complex [13$_{Ca}$][I$_2$] selectively produces cis-CHC from CO_2 and CHO. The influence of the nucleophilic anions onto the catalytic reaction was determined. The catalytic activity and selectivity of cis-CHC formation increases in the order OTf$^-$ > Cl$^-$ > Br$^-$ > I$^-$. Since iodide exhibited the highest activity, further experiments were performed with complexes featuring iodide as counter anion. Optimization of the reaction conditions increases the catalytic activity of [13$_{Ca}$][I$_2$] from a TOF of 17 h^{-1} to 34 h^{-1}, while retaining the high cis-selectivity. The influence of the central cation onto the reaction was determined under optimized reaction conditions. The catalytic activity decreases with the higher alkaline earth metals $Ca^{2+} \approx Sr^{2+} > Ba^{2+}$, which is rationalized by the lower solubility of the barium complex [13$_{Ba}$][I$_2$]. Various epoxide substrates featuring different functional groups were tested to determine the substrate scope. Most COCs were obtained at high selectivities and conversions. Epoxides featuring unsaturated substituents lead to increased side product formation, albeit at high conversions. Substrates featuring sterically demanding substituents close to the oxirane ring are not quantitatively converted to the corresponding COCs. [13$_{Ca}$][I$_2$] also exhibits high catalytic activity and regioselectivity in the formation of oxazolidinone from PO with the heterocumulene phenyl isocyanate. The catalytic activity with a TOF value of 1280 h^{-1} surpasses other mononuclear vanadium systems with a TOF value of 8 h^{-1}.[43]

Further studies should focus on the high catalytic activity of [13$_{Ca}$][I$_2$] regarding the formation of oxazolidinones from epoxides and isocyanates. Further studies should also examine the influence of the chiral ligand environment onto the performed reactions to determine whether chiral induction occurs to produce enantiomerically enriched products.

Catalytic Transformations of Heterocumulenes with Epoxides

B.4.4 Experimental

B.4.4.1 General Considerations

The catalysts used in this chapter were synthesized according to the procedures reported in chapter B.2 and B.3. CHO and PO were dried over CaH_2, distilled and degassed. Phenyl isocyanate was dried over P_4O_{10}, distilled and degassed. All other chemicals were used without further purification.

B.4.4.2 Synthesis of Trivanadyl Complexes

[((R,R)-^{C2Ph2}L)(VO)$_3$SrI$_2$] ([13$_{Sr}$][I$_2$])

[13$_{Sr}$][I$_2$] was prepared analogously to [13$_{Ca}$][(OTf)$_2$] with [9$_{Sr}$][I$_2$] (100 mg, 73.1 µmol, 1.00 equiv.) and VO(acac)$_2$ (58.1 mg, 219 µmol, 3.00 equiv.) in MeOH/MeCN (1:1). The product [13$_{Sr}$][I$_2$] was obtained as a dark red/orange solid (102 mg, 65.4 µmol, 90%). **FAB MS** (pos. mode, 3-NBA matrix): m/z = 1308.5 ([((R,R)-^{C2Ph2}L)(VO)$_3$Sr]$^+$), 1243.6 ([(H(R,R)-^{C2Ph2}L)(VO)$_2$Sr]$^+$), 1460.5 ([((R,R)-^{C2Ph2}L)(VO)$_3$Sr + 3-NBA]$^+$). **IR** in KBr: ($\tilde{v}$, cm^{-1}) 3417 (w), 3058 (w), 3026 (w), 2923 (w), 2873 (w), 1609 (vs, C=N), 1533 (w), 1504 (m), 1465 (m), 1444 (s), 1384 (s), 1325 (vs), 1268 (w), 1232 (s), 1180 (m), 1120 (w), 1046 (s), 1002 (s, V=O), 918 (vw), 884 (vw), 868 (vw), 838 (vw), 753 (m), 701 (m), 641 (w), 618 (vw), 609 (vw), 573 (w), 546 (w), 478 (w).

[((R,R)-^{C2Ph2}L)(VO)$_3$BaI$_2$] ([13$_{Ba}$][I$_2$])

[13$_{Ba}$][I$_2$] was prepared analogously to [13$_{Ca}$][(OTf)$_2$] with [9$_{Ba}$][I$_2$] (110 mg, 77.6 µmol, 1.00 equiv.) and VO(acac)$_2$ (61.7 mg, 233 µmol, 3.00 equiv.) in MeOH/MeCN (1:1). The product [13$_{Ba}$][I$_2$] was obtained as a dark red/orange solid (104 mg, 64.5 µmol, 83%). **FAB MS** (pos. mode, 3-NBA matrix): m/z = 1358.5 ([((R,R)-^{C2Ph2}L)(VO)$_3$Ba]$^+$), 1292.6 ([H((R,R)-^{C2Ph2}L)(VO)$_2$Ba]$^+$), 1485.4 ([((R,R)-^{C2Ph2}L)(VO)$_3$BaI]$^+$), 1510.5 ([((R,R)-^{C2Ph2}L)(VO)$_3$Ba + 3-NBA]$^+$). **IR** in KBr: ($\tilde{v}$, cm^{-1}) 3416 (m), 3059 (w), 3029 (w), 2919 (w), 1613 (vs, C=N), 1596 (w), 1496 (w), 1443 (m), 1384 (vs), 1323 (s), 1231 (m), 1177 (m), 1120 (w), 1054 (w), 1030 (vw), 989 (m, V=O), 866 (vw), 839 (vw), 752 (m), 703 (m), 688 (w), 618 (w), 570 (w), 544 (w), 474 (w). **Elemental Analysis:** Calcd. for $C_{66}H_{48}BaI_2N_6O_6V_3(Et_2O)_2$: C 50.46, H 3.89, N 4.77; found: C 50.48, H 3.97, N 4.92.

B.4.4.3 Catalytic Reactions of CO_2 with Epoxides

Catalytic reactions of epoxide with CO_2 at elevated pressures were performed in nickel alloy autoclaves equipped with a magnetic stirrer bar. The bottom part of the autoclave and the stirrer bar were heated overnight in a drying oven at 130 °C and introduced hot into the glovebox. Once cooled to room temperature, the autoclave was assembled inside of the glovebox.

General procedure for the catalytic formation of cyclic carbonates:
The epoxide was added to the catalyst in a glass vial inside of the glovebox. The reaction mixture was transferred into the autoclave which was equipped with a magnetic stirrer bar. The autoclave was assembled inside of the glovebox. Outside of the glovebox, the connector and the autoclave wear flushed each three times with CO_2 by applying a pressure of 5 bar and releasing the pressure again. The designated pressure was applied, and the autoclave heated inside of an oil bath at the desired temperature while stirring. After 15 min the pressure was adjusted to the exact pressure in case the pressure was lower than the desired one. After completion of the reaction time, the autoclave was cooled inside of an ice/water bath to room temperature and vented. The reaction mixture was collected. The conversion and selectivities were determined by NMR spectroscopy in chloroform-d_1.[51, 52]

General procedure for the catalytic co-polymerization of CHO with CO_2:
The epoxide was added to the catalyst in a glass vial inside of the glovebox. The reaction mixture was transferred into the autoclave which was equipped with a magnetic stirrer bar. The autoclave was assembled inside of the glovebox. Outside of the glovebox, the connector and the autoclave wear flushed each three times with CO_2 by applying a pressure of 5 bar and releasing the pressure again. A CO_2 pressure of 8 bar was applied, and the autoclave heated inside of an oil bath at 100 °C while stirring. After 15 min the pressure was adjusted to 10 bar. After completion of the reaction time, the autoclave was cooled inside of an ice/water bath to room temperature and vented. A sample was taken from the reaction mixture and analyzed by NMR spectroscopy. The reaction mixture was taken up in DCM and precipitated through fast addition of the solution to methanol. The precipitated polymer was collected and dried under reduced pressure to afford a light-yellow solid. The polymer was analyzed according to literature procedures by NMR, IR spectroscopy and SEC.[46-48]

B.4.4.4 Catalytic Reactions of Phenyl Isocyanate with PO

Phenyl isocyanate was added to a solution of **[13$_{Ca}$][I$_2$]** (0.1 mol%) in PO inside of a flask equipped with a J. Young-type spindle valve under inert gas conditions. The reaction mixture was stirred for the specified time at 80 °C. After completion of the reaction time, a sample was analyzed by NMR spectroscopy in chloroform-d_1. The conversion and selectivity were determined according to literature procedures by NMR spectroscopy.[41]

B.4.5 References

1. Shaikh, R. R.; Pornpraprom, S.; D'Elia, V. *ACS Catal.* **2017**, *8*, 419-450.
2. Büttner, H.; Longwitz, L.; Steinbauer, J.; Wulf, C.; Werner, T. *Top. Curr. Chem.* **2017**, *375*, 50.
3. Poland, S. J.; Darensbourg, D. J. *Green Chem.* **2017**, *19*, 4990-5011.
4. Müller, T. E.; Gürtler, C.; Pohl, M.; Subhani, M. A.; Leitner, W. Method for Producing Polyether Carbonate Polyols. US 2018/0291148 A1, 2018.
5. Hofmann, A.; Migeot, M.; Thißen, E.; Schulz, M.; Heinzmann, R.; Indris, S.; Bergfeldt, T.; Lei, B.; Ziebert, C.; Hanemann, T. *ChemSusChem* **2015**, *8*, 1892-1900.
6. Vivek, J. P.; Berry, N.; Papageorgiou, G.; Nichols, R. J.; Hardwick, L. J. *J. Am. Chem. Soc.* **2016**, *138*, 3745-3751.
7. Schäffner, B.; Schäffner, F.; Verevkin, S. P.; Börner, A. *Chem. Rev.* **2010**, *110*, 4554-4581.
8. Barbachyn, M. R.; Ford, C. W. *Angew. Chem. Int. Ed.* **2003**, *42*, 2010-2023.
9. Roush, W. R.; James, R. A. *Aust. J. Chem.* **2002**, *55*, 141-146.
10. Zhu, Y.; Romain, C.; Williams, C. K. *Nature* **2016**, *540*, 354-362.
11. Inoue, S.; Koinuma, H.; Tsuruta, T. *J. Polym. Sci.Pol. Lett.* **1969**, *7*, 287-292.
12. Trott, G.; Saini, P. K.; Williams, C. K. *Phil. Trans. R. Soc. A* **2016**, *374*, 20150085.
13. Darensbourg, D. J.; Holtcamp, M. W. *Coord. Chem. Rev.* **1996**, *153*, 155-174.
14. Chatterjee, C.; Chisholm, M. H. *Inorg. Chem.* **2011**, *50*, 4481-4492.
15. Chatterjee, C.; Chisholm, M. H. *Inorg. Chem.* **2012**, *51*, 12041-12052.
16. Chatterjee, C.; Chisholm, M. H.; El-Khaldy, A.; McIntosh, R. D.; Miller, J. T.; Wu, T. *Inorg. Chem.* **2013**, *52*, 4547-4553.
17. Aida, T.; Inoue, S. *J. Am. Chem. Soc.* **1983**, *105*, 1304-1309.
18. Darensbourg, D. J. *Chem. Rev.* **2007**, *107*, 2388-2410.
19. S, S.; Min, J. K.; Seong, J. E.; Na, S. J.; Lee, B. Y. *Angew. Chem. Int. Ed.* **2008**, *47*, 7306-7309.
20. Darensbourg, D. J.; Yarbrough, J. C.; Ortiz, C.; Fang, C. C. *J. Am. Chem. Soc.* **2003**, *125*, 7586-7591.
21. Schütze, M.; Dechert, S.; Meyer, F. *Chem. Eur. J.* **2017**, *23*, 16472-16475.
22. Deacy, A. C.; Durr, C. B.; Garden, J. A.; White, A. J. P.; Williams, C. K. *Inorg. Chem.* **2018**, *57*, 15575-15583.
23. Jutz, F.; Buchard, A.; Kember, M. R.; Fredriksen, S. B.; Williams, C. K. *J. Am. Chem. Soc.* **2011**, *133*, 17395-405.
24. Jiang, H.; Qi, C.; Wang, Z.; Zou, B.; Yang, S. *Synlett* **2007**, *2*, 255-258.
25. Büttner, H.; Steinbauer, J.; Wulf, C.; Dindaroglu, M.; Schmalz, H. G.; Werner, T. *ChemSusChem* **2017**, *10*, 1076-1079.
26. Steinbauer, J.; Spannenberg, A.; Werner, T. *Green Chem.* **2017**, *19*, 3769-3779.
27. Gao, P.; Zhao, Z.; Chen, L.; Yuan, D.; Yao, Y. *Organometallics* **2016**, *35*, 1707-1712.
28. Della Monica, F.; Maity, B.; Pehl, T.; Buonerba, A.; De Nisi, A.; Monari, M.; Grassi, A.; Rieger, B.; Cavallo, L.; Capacchione, C. *ACS Catal.* **2018**, *8*, 6882-6893.
29. Wu, X.; Chen, C.; Guo, Z.; North, M.; Whitwood, A. C. *ACS Catal.* **2019**, *9*, 1895-1906.
30. Longwitz, L.; Steinbauer, J.; Spannenberg, A.; Werner, T. *ACS Catal.* **2018**, *8*, 665-672.
31. Yang, Y.; Hayashi, Y.; Fujii, Y.; Nagano, T.; Kita, Y.; Ohshima, T.; Okuda, J.; Mashima, K. *Catal. Sci. Technol.* **2012**, *2*, 509-513.
32. D'Elia, V.; Pelletier, J. D. A.; Basset, J.-M. *ChemCatChem* **2015**, *7*, 1906-1917.
33. Coletti, A.; Whiteoak, C. J.; Conte, V.; Kleij, A. W. *ChemCatChem* **2012**, *4*, 1190-1196.
34. Liu, X.; Wang, M. Y.; Wang, S. Y.; Wang, Q.; He, L. N. *ChemSusChem* **2017**, *10*, 1210-1216.
35. Seo, U. R.; Chung, Y. K. *Green Chem.* **2017**, *19*, 803-808.
36. Xu, B.; Wang, P.; Lv, M.; Yuan, D.; Yao, Y. *ChemCatChem* **2016**, *8*, 2466-2471.
37. Speranza, G. P.; Peppel, W. J. *J. Org. Chem.* **1958**, *23*, 1922-1924.
38. Quian, C.; Zhu, D. *Synlett* **1994**, *0*, 129-130.

39. Herweh, J. E.; Foglia, T. A.; Swern, D. *J. Org. Chem.* **1968**, *33*, 4029-4033.
40. Zhang, X.; Chen, W.; Zhao, C.; Li, C.; Wu, X.; Chen, W. Z. *Synth. Commun.* **2010**, *40*, 3654-3659.
41. Fujiwara, M.; Baba, A.; Matsuda, H. *J. Hetrocyclic Chem.* **1988**, *25*, 1351-1357.
42. Shibata, I.; Baba, A.; Iwasaki, H.; Matsuda, H. *J. Org. Cem.* **1986**, *27*, 77-80.
43. Beattie, C.; North, M. *RSC Adv.* **2014**, *4*, 31345-31352.
44. Kleemann, J. Homogenkatalysatoren für die Copolymerisation von Cyclohexenoxid mit CO_2 auf Basis multimetallischer Komplexe von Seltenerdmetallen und Zink. Doctoral Thesis, RWTH Aachen University, Aachen, 2017.
45. Nagae, H.; Aoki, R.; Akutagawa, S. N.; Kleemann, J.; Tagawa, R.; Schindler, T.; Choi, G.; Spaniol, T. P.; Tsurugi, H.; Okuda, J.; Mashima, K. *Angew. Chem. Int. Ed.* **2018**, *57*, 2492-2496.
46. Lednor, P. W.; Rol, N. C. *J. Chem. Soc., Chem. Commun.* **1985**, *0*, 598-599.
47. Schilling, F. C.; Tonelli, A. E. *Macromolecules* **1986**, *19*, 1337-1343.
48. Kuran, W.; Listoś, T. *Macromol. Chem. Phys.* **1994**, *195*, 977-984.
49. Trummal, A.; Lipping, L.; Kaljurand, I.; Koppel, I. A.; Leito, I. *J. Phys. Chem. A* **2016**, *120*, 3663-3669.
50. Xiao, Y.; Wang, Z.; Ding, K. *Macromolecules* **2006**, *39*, 128-137.
51. Miceli, C.; Rintjema, J.; Martin, E.; Escudero-Adán, E. C.; Zonta, C.; Licini, G.; Kleij, A. W. *ACS Catal.* **2017**, *7*, 2367-2373.
52. Lu, J.; Toy, P. *Synlett* **2011**, *5*, 659-662.

C. Summary

Whereas tetradentate (OSSO)-type ligands mostly allow for the preparation of mononuclear transition metal complexes,[1-6] ligands that feature two or more of the structurally related (ONNO)-type subunits may give access to multinuclear heterometallic complexes.[7, 8] Comparative studies of molybdenum complexes in the biologically relevant oxidation states +IV, +V and +VI stabilized by tetradentate (OSSO)-type ligands led to the development of a highly active oxygen atom transfer (OAT) catalyst.[9-11] Detailed studies on complexes featuring tris(ONNO)-type ligands revealed a new approach toward heterometallic $3d$ transition metal complexes that resolves the limitations of core/shell template directed synthesis and avoids undesired side reactions during ligand synthesis.[7, 12, 13] The heterometallic complexes exhibit metal cooperative effects resulting in magnetic interactions of the paramagnetic centers and in enhanced catalytic activity in transformations of heterocumulenes with epoxides. This thesis further contributes to the field of mono- and tetranuclear metal complexes stabilized by (OSSO)- or tris(ONNO)-type ligands. Based on these ligands, the structural, magnetic and electrochemical properties of various complexes are reported.

Chapter B.1 describes the synthesis and characterization of mononuclear molybdenum and tungsten complexes featuring (OSSO)-type ligands. The synthesis of mono(oxo) molybdenum(V) and tungsten(V) complexes was reported and the structural properties assessed. The electronic and magnetic properties of dichloro molybdenum(IV), mono-oxo molybdenum(V) and dioxo molybdenum(VI) complexes were determined. Due to the hard bis(phenolato) ligands the reductions and oxidations occur at more negative and positive potentials than complexes featuring the analogous soft bis(thiophenolato) ligands. Comparative studies indicate a large influence of the initial oxidation states of the catalysts on OAT reactions. The dichloro molybdenum complex exhibits the highest catalytic activity due to formation of a highly reactive molybdenum(VI) intermediate. The catalytic activity is even surpassed by the tetrachloro complex [MoCl$_4$(MeCN)$_2$] with a TOF value of 1590 h^{-1}.

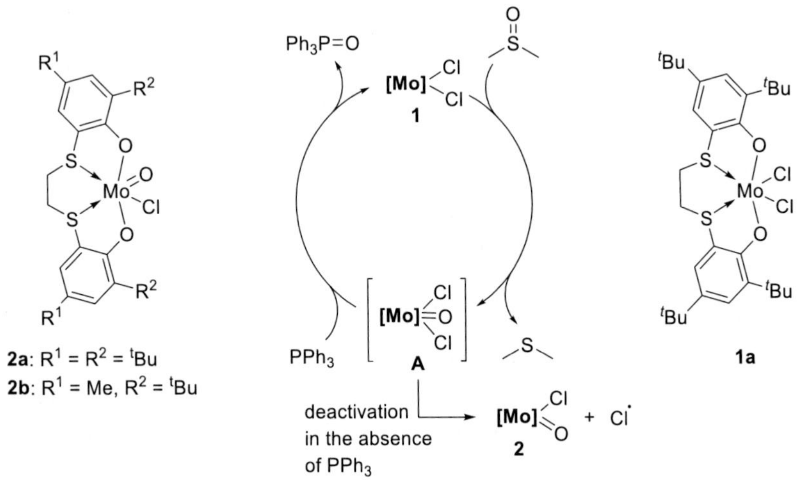

Summary

Scheme B.4.1. Oxygen atom transfer reactions from DMSO to PPh$_3$ catalyzed by a dichloro molybdenum(IV) complex which decomposes to a mono-oxo molybdenum(V) complex in the absence of an oxygen atom scavenger.

In chapter B.2 the monometallic template directed synthesis of tris(ONNO)-type macrocyclic ligands based on 3,6-diformylcatechol was reported (Scheme B.4.2). The mononuclear complexes feature vacant salen-type binding sites and an occupied central 18c6-type cavity. Pre-coordination of the metal templates to 3,6-diformylcatechol and subsequent condensation with diamines affords mononuclear calcium, strontium, barium, lanthanum and cerium(III) complexes with triflate, tosylate, chloride, bromide or iodide counter anions. The monometallic template synthesis is applicable to C$_2$-bridging units such as 1,2-cyclohexanediamine or 1,2-diphenylethylenediamine. The macrocyclic ligand is sufficiently stabilized by coordination of the 18c6-type cavity to a Lewis acidic metal center to inhibit Schiff base rearrangements, which often occur in protonated macrocyclic pro-ligands with more flexible bridging units.

Summary

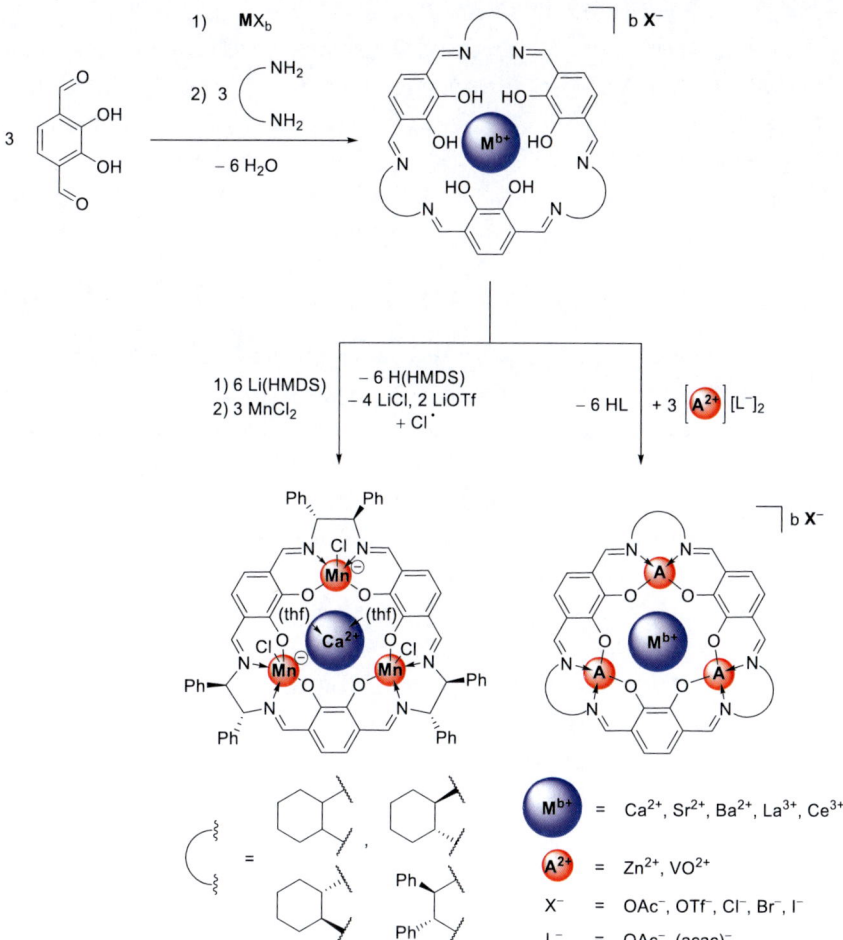

Scheme B.4.2. Synthesis of mononuclear and tetranuclear heterometallic complexes featuring tris(ONNO)-type ligands.

The monometallic complexes are thus suitable for preparation of heterometallic tetranuclear complexes through subsequent metalation of the vacant salen-type binding sites with transition metals. This approach was utilized in chapter B.3 to prepare heterometallic trizinc, trivanadyl and trimanganese complexes featuring tris(ONNO)-type ligands. Metalation of the salen-type binding sites was achieved through protonolysis of the mononuclear complexes with $Zn(OAc)_2$ or $VO(acac)_2$, or through deprotonation with Li(HMDS) and subsequent salt metathesis with $MnCl_2$ (Scheme B.4.2). Antiferromagnetic exchange interactions are observed in the case of the calcium trivanadyl complex [13$_{Ca}$][(OTf)$_2$] and calcium trimanganese complex [16$_{Ca}$][Cl$_3$]

Summary

due to spin coupling between the paramagnetic centers. The calcium trivanadyl complex [13$_{Ca}$][(OTf)$_2$] is another example of a frustrated spin system. Molecular structures of the calcium trizinc [11$_{Ca}$][(OAc)$_2$] and trimanganese [16$_{Ca}$][Cl$_3$] complexes confirm exclusive coordination of the transition metal centers in the salen-type binding sites. The calcium trimanganese complex [16$_{Ca}$][Cl$_3$] features one manganese(III) and two manganese(II) centers. Unlike the calcium trivanadyl complex, the spin of each manganese center of [16$_{Ca}$][Cl$_3$] is larger than 1/2, allowing for further coupling scenarios that preclude description as a frustrated spin system. [16$_{Ca}$][Cl$_3$] features a [Mn$_3$Ca] core that may allow comparative studies with the [Mn$_3$Ca] cubane found in the natural oxygen evolving complex in photosystem II.

Chapter B.4 describes the application of mononuclear and tetranuclear complexes synthesized in sections B.2 and B.3 in catalytic transformations of heterocumulenes with epoxides. The tested mononuclear complexes are catalytically inactive, whereas the heterometallic complexes exhibit catalytic activity. Due to metallic cooperativity, the calcium trizinc and calcium trimanganese complexes convert CO_2 and CHO to poly(cyclohexene carbonate), whereas the trivanadyl complexes produce cyclic carbonates. The reaction conditions of the CHC formation were optimized using trivanadyl catalysts. The influence of the central cation and the anions was determined. The catalytic activity and selectivity of *cis*-CHC formation increases in the order OTf$^-$ < Cl$^-$ < Br$^-$ < I$^-$, as I$^-$ is a better leaving group than the other anions. The catalytic activity decreases with the higher alkaline earth metals $Ca^{2+} \approx Sr^{2+} > Ba^{2+}$, due to lower solubility of the complexes. The optimal catalyst [13$_{Ca}$][I$_2$] selectively converts CO_2 and epoxides to COCs. Surprisingly, the catalyst exhibits high catalytic activity and regioselectivity in the formation of oxazolidinone from propylene oxide and phenyl isocyanate with a TOF of 1280 h^{-1}.

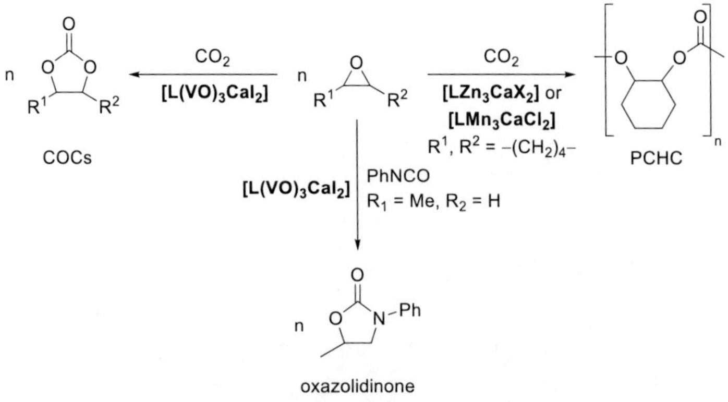

Scheme B.4.3. Catalytic transformations of heterocumulenes with epoxides.

D. Appendix
D.1 General Experimental Procedures

All procedures were performed under an argon atmosphere using standard Schlenk-line or glovebox techniques unless otherwise stated. All glass equipment was dried at 130 °C in an oven overnight and assembled inside of a glovebox or flame dried at the Schlenk-line. Tetrahydrofuran (THF), toluene, dichloromethane (DCM), diethyl ether (Et_2O) and n-pentane were dried with a MBraun SPS and stored over molecular sieves. All other solvents were dried, vacuum transferred and degassed by standard methods and stored over molecular sieves. All other chemicals were purchased from commercial sources and used as received without further purification.

NMR measurements were performed on a Bruker Avance II 400 or a BRUKER AVANCEIII-400 spectrometer at 298 K unless otherwise specified. ^1H and ^{13}C NMR spectra were referenced to residual proton signals of the deuterated solvents and reported relative to tetramethylsilane.[14] The residual proton signals of MeCN-d_2 and MeOD-d_3 in MeCN-d_3/MeOD-d_4 (1:1) were determined at ^1H NMR: δ 3.277, 1.975; and at $^{13}C\{^1H\}$ NMR: δ 118.45, 49.07, 0.91. Chemical shifts (δ) are reported in parts per million (ppm). Standard abbreviations indicating multiplicities were used as follows: s (singlet), d (doublet), t (triplet) m. (multiplet) and br. (broad). Assignment of the resonances in the ^1H and $^{13}C\{^1H\}$ NMR spectra was performed with multidimensional NMR spectroscopy (NOESY, COESY, HSQC, HMBC). ^{13}C and ^{19}F spectra were recorded broadband ^1H decoupled.

Samples for IR spectroscopy were processed with dried KBr to pellets. IR spectra were recorded on KBr pellets using an AVATAR 360 FT-IR or a JASCO FT/IR-4200 spectrometer. Abbreviations for IR spectra were used as follows: vw (very weak), w (weak), m (medium), s (strong), vs (very strong), br (broad), sh (shoulder).

UV/Vis spectra were recorded with a Shimadzu UV-2600 spectrometer. The compounds were dissolved in equivolumic solutions of MeOH and MeCN.

Elemental analyses were performed on an Elementar Analysesysteme GmbH Elementar EL device by the Institute of Organic Chemistry at the RWTH Aachen University. A lower carbon content results from incomplete combustion as has been reported for other metal containing coordination compounds.[15] The chloride content was determined by ion chromatography on a Metrohm 792 Basic IC equipped with a METROSEP A Supp 4 (6.1006.430, 4.0 x 250 mm) column and a suppressor unit, using a buffer solution of Na_2CO_3 (1.8 mM) and $NaHCO_3$ (1.7 mM) at a flow rate of 1.00 mL min^{-1}. Calibration was performed with aqueous samples of $MnCl_2 \cdot 4\ H_2O$ and H_2SO_4 in ultrapure water. Samples were decomposed by sonication of the solids in H_2SO_4/ultrapure water. The resulting suspensions were extracted with ultrapure water

Appendix

(< 0.1 μS cm^{-1}), filtered through a PTFE filter (0.45 μm) and diluted to 100 mL. The content of manganese and calcium were determined on a Perkin Elmer AAS model AAnalyst20 by the company Mikroanalytisches Labor Kolbe.

FAB mass spectra were recorded on a JMS-700 mass spectrometer by the mass spectrometry department of the Graduate School of Engineering Science at the Osaka University or on a Finnigan MAT 95 mass spectrometer by the mass spectrometry department of the Institute of Inorganic Chemistry at the RWTH Aachen University. ESI mass spectra were recorded on a ThermoFinnigan LCQ Deca XP plus mass spectrometer by the mass spectrometry department of the Institute of Organic Chemistry at the RWTH Aachen University.

Molecular weights and polydispersity indices (PDI) were determined by Monika Paul by size exclusion chromatography (SEC) in THF at 298 K, at a flow rate of 1 mL min^{-1} utilizing an PSS SECurity HPLC pump, an Agilent 1100 Series reflective index detector and 8 x 600 mm, 8 x 300 mm, 8 x 50 mm PSS SDV linear M columns. Calibration standards were commercially available narrowly distributed linear polystyrene samples that cover a broad range of molar masses ($10^3 < M_n < 2 \cdot 10^6$ g mol^{-1}).

Electrochemical characterization was performed with a Metrohm Autolab PGSTAT101 using a Pt-disk working electrode, a Pt-wire counter electrode and a Ag-wire quasi-reference electrode inside of a glovebox. The voltammograms were referenced to the Fc/Fc$^+$ couple. Fresh distilled and degassed solvents were used for performing the electrochemical characterization.

EPR measurements were performed together with Kristina Keisers on a Magnettech MS 400 X-Band spectrometer (B_0-filed: 50 – 450 mT) at the indicated temperatures in the group of Prof. Sonja Herres-Pawlis. Simulations and fittings of the experimental EPR spectra were performed with the EasySpin software (v. 5.2.12).[16, 17] Fittings of the experimental EPR spectra were performed with a Nelder/Mead downhill simplex optimization algorithm.[18, 19]

The magnetic data of [13$_{Ca}$][(OTf)$_2$] and of [16$_{Ca}$][Cl$_3$] were recorded by the group of Prof. Paul Kögerler using a Quantum Design MPMS-5XL SQUID magnetometer. The polycrystalline samples were compacted and immobilized into a cylindrical PTFE capsule. The data were acquired as a function of the magnetic field (0 – 5.0 T at 2.0 K) and the temperature (2.0 – 290 K at 0.1 T). The data of [13$_{Ca}$][(OTf)$_2$] and [16$_{Ca}$][Cl$_3$] were corrected for the diamagnetic contributions of the sample holder and the compound χ_{dia} = -7.80×10^{-4} cm^3 mol^{-1} ([13$_{Ca}$][(OTf)$_2$]) and χ_{dia} = -7.38×10^{-4} cm^3 mol^{-1} ([16$_{Ca}$][Cl$_3$]). Since the data indicated minor ferromagnetic impurities, they were additionally corrected according to standard procedures.[20-22] Simulation and evaluation of the data were performed by Dr. Dr. Jan van Leusen.

Appendix

Magnetic data of compounds in solution were obtained by Evans method on a Bruker Avance II 400 or a Bruker Avance III 400 spectrometer at 298 K.[23, 24]

Diffraction data of **2b** were collected by the group of Prof. Ulli Englert at the RWTH Aachen University on a Bruker D8 goniometer with Mo K_α radiation with an APEX CCD area-detector at 100 K in ω-scan mode. Data reductions and absorption corrections were carried out with the programs SAINT and SADABS by the Englert groupt.[25, 26] The structure was solved with SIR-92.[27] All refinements were performed against F^2 with SHELXL-2013[28] as implemented in the program system WinGX.[29]

Diffraction data of **[11$_{Ca}$(dmso)$_2$(H$_2$O)(OAc)$_2$]** and **[16$_{Ca}$][Cl$_3$]** were collected at 100 K on a Eulerian 4-circle diffractometer (STOE STADIVARI) with Cu K_α radiation (graded multilayer mirror, λ = 1.54186 Å) using ω scans. The programs X-Area[30-32] and STOE X-Red32[33] were used for data reduction and absorption correction. Collection of the diffraction data of **[11$_{Ca}$(dmso)$_2$(H$_2$O)(OAc)$_2$]** and **[16$_{Ca}$][Cl$_3$]**, data reduction and absorption correction were performed by Florian Ritter.

Diffraction data of **[9$_{Ca}$(MeOH)$_2$][(OTf)$_2$]** were collected with assistance of Ass. Prof. Haruki Nagae at the Graduate School of Engineering Science at the Osaka University on a Rigaku XtaLAB P200 diffractometer with Mo K_α radiation at 113 K in ω-scan mode. Data reductions and absorption corrections were carried out with the program CrystalClear (Rigaku).[34, 35] The structure was solved with SIR-92.[27] All refinements were performed against F^2 with SHELXL-2013[28] as implemented in the program system WinGX.[29] The crystal packing of **[9$_{Ca}$(MeOH)$_2$][(OTf)$_2$]** and **[16$_{Ca}$][Cl$_3$]** contain large voids due to disordered solvent molecules which were removed by using the routine SQUEEZE as implemented in the program system PLATON.[36]

The solution and refinement of all crystal structures was performed by Dr. Thomas P. Spaniol. The graphical representations were prepared with the DIAMOND software.[37]

Appendix

D.2 Crystal Structure Parameters

	2b	[9$_{Ca}$(MeOH)$_2$][(OTf)$_2$]	[11$_{Ca}$(dmso)$_2$(H$_2$O)(OAc)$_2$]
Chemical formula	C$_{24}$H$_{32}$ClMoO$_3$S$_2$	C$_{70}$H$_{56}$CaF$_6$N$_6$O$_{14}$S$_2$	Ca$_2$Zn$_6$S$_{20}$O$_{43}$N$_{12}$C$_{180}$
Mw [g mol^{-1}]	564.00	1423.40	4115.851
cryst. size [mm]	0.07 x 0.45 x 0.47	0.07 x 0.11 x 0.12	
color	(plate)	(block)	(block)
	dark red	red	orange
space group	$P\,2_1/c$	$C\,2$	$P\,3_1$
a [Å]	10.012(2)	27.744(3)	18.024(3)
b [Å]	12.807(3)	16.1465(18)	18.024(3)
c [Å]	19.661(4)	27.970(4)	54.630(11)
α [°]	90	90	90
β [°]	90.076(3)	118.692(10)	90
γ [°]	90	90	120
V [Å^3]	2521.0(9)	10991(2)	15369.7(50)
Z	4	6	3
T [K]	100(2)	113(2)	100(2)
μ(Mo K_α) [mm^{-1}]	0.815	0.222	
μ(Cu K_α) [mm^{-1}]	/	/	3.726
$d_{calc.}$	1.486	1.290	1.406
F(000)	1164	4416	6780
Reflns	31102	78574	34501
R_{int}	0.0611	0.0632	0.0496
Independent reflns	5452	23981	21641
Observed reflns	5206	15554	16239
$R1$, $wR2$ [$I \geq 2\sigma(I)$]	0.0424, 0.0897	0.0599, 0.1270	0.1185, 0.3178
$R1$, $wR2$ (all data)	0.0452, 0.0911	0.1107, 0.1455	0.1373, 0.3293
GOF (F^2)	1.137	1.013	0.992

	$[16_{Ca}][Cl_3]$
Chemical formula	$C_{182}H_{175}Ca_2Cl_6Mn_6N_{15}O_{23}$
Mw [g mol^{-1}]	3562.86
cryst. size [mm]	0.260 x 0.220 x 0.200
color	(block) brown
space group	$P2_1$
a [Å]	14.290(3)
b [Å]	10.714(2)
c [Å]	30.634(6)
$α$ [°]	90
$β$ [°]	93.89(3)
$γ$ [°]	90
V [Å^3]	4679.4(16)
Z	1
T [K]	100(2)
$μ$(Mo $K_α$) [mm^{-1}]	/
$μ$(Cu $K_α$) [mm^{-1}]	4.994
$d_{calc.}$	1.264
F(000)	1848
Reflns	27557
R_{int}	0.0300
Independent reflns	14245
Observed reflns	10746
$R1$, $wR2$ [$I ≥ 2σ(I)$]	0.0502, 0.1223
$R1$, $wR2$ (all data)	0.0658, 0.1277
GOF (F^2)	0.951

D.3 References

1. Capacchione, C.; Proto, A.; Ebeling, H.; Mülhaupt, R.; Möller, K.; Spaniol, T. P.; Okuda, J. *J. Am. Chem. Soc.* **2003**, *125*, 4964-4965.
2. Beckerle, K.; Capacchione, C.; Ebeling, H.; Manivannan, R.; Mülhaupt, R.; Proto, A.; Spaniol, T. P.; Okuda, J. *J. Organomet. Chem.* **2004**, *689*, 4636-4641.
3. Capacchione, C.; Manivannan, R.; Barone, M.; Beckerle, K.; Centore, R.; Oliva, L.; Proto, A.; Tuzi, A.; Spaniol, T. P.; Okuda, J. *Organometallics* **2005**, *24*, 2971-2982.
4. Ma, H.; Spaniol, T. P.; Okuda, J. *Inorg. Chem.* **2008**, *47*, 3328-3339.
5. Meppelder, G.-J. M.; Beckerle, K.; Manivannan, R.; Lian, B.; Raabe, G.; Spaniol, T. P.; Okuda, J. *Chem. Asian J.* **2008**, *3*, 1312-1323.
6. Kapelski, A.; Buffet, J. C.; Spaniol, T. P.; Okuda, J. *Chem. Asian J.* **2012**, *7*, 1320-1330.
7. Akine, S.; Nabeshima, T. *Dalton Trans.* **2009**, *47*, 10395-10408.
8. Clarke, R. M.; Storr, T. *Dalton Trans.* **2014**, *43*, 9380-9391.
9. Schindler, T.; Sauer, A.; Spaniol, T. P.; Okuda, J. *Organometallics* **2018**, *37*, 4336-4340.
10. Sauer, A. Early Transition Metal Bis(phenolate) Complexes for Selective Catalysis. Doctoral Thesis, RWTH Aachen University, Aachen, 2013.
11. Beckerle, K.; Sauer, A.; Spaniol, T. P.; Okuda, J. *Polyhedron* **2016**, *116*, 105-110.
12. Akine, S.; Sunaga, S.; Taniguchi, T.; Miyazaki, H.; Nabeshima, T. *Inorg. Chem.* **2007**, *46*, 2959-2961.
13. MacLachlan, M. J. *Pure Appl. Chem.* **2006**, *78*, 873-888.
14. Fulmer, G. R.; Miller, A. J. M.; Sherden, N. H.; Gottlieb, H. E.; Nudelman, A.; Stoltz, B. M.; Bercaw, J. E.; Goldberg, K. I. *Organometallics* **2010**, *29*, 2176-2179.
15. Marcó, A.; Compañó, R.; Rubio, R.; Casals, I. *Microchim. Acta* **2003**, *142*, 13-19.
16. Stoll, S.; Schweiger, A. *J. Magn. Reson.* **2006**, *178*, 42-55.
17. *MATLAB and Statistics Toolbox Release 2016b*, The MathWorks, Inc.: Natick, Massachusetts, USA, 2016.
18. Nelder, J. A.; Mead, R. *Comput. J.* **1965**, *7*, 308-313.
19. Nelder, J. A.; Mead, R. *Comput. J.* **1965**, *8*, 27.
20. Lueken, H., *Magnetochemie*. Vieweg+Teubner Verlag: Stuttgart, 1999.
21. Honda, K. *Ann. Phys.* **1910**, *337*, 1027-1063.
22. Kohlrausch, F., *Praktische Physik*. 23 ed.; Teubner: Stuttgart, 1985; Vol. 2.
23. Evans, D. F. *J. Chem. Soc.* **1959**, *0*, 2003-2005.
24. Evans, D. F.; Fazakerley, G. V.; Phillips, R. F. *J. Chem. Soc. A* **1971**, *0*, 1931-1934.
25. Bruker *SAINT-Plus*, Bruker AXS Inc.: Madison, Wisconsin, USA, 1999.
26. Bruker *SADABS*, Bruker AXS Inc.: Madison, Wisconsin, USA, 2004.
27. Altomare, A.; Cascarano, G.; Giacovazzo, C.; Guagliardi, A. *J. Appl. Cryst.* **1993**, *26*, 343-350.
28. Sheldrick, G. M. *Acta Cryst. A* **2008**, *64*, 112-122.
29. Farrugia, L. J. *J. Appl. Cryst.* **2012**, *45*, 849-854.
30. *X-Area Recipe*, 1.33.0.0; STOE: 2015.
31. *X-Area Integrate*, 1.71.0.0; STOE: 2016.
32. *X-Area LANA*, 1.71.4.0; STOE: 2017.
33. STOE X-Red32, absorption correction by Gaussian integration, analogous to P. Coppens, The Evaluation of Absorption and Extinction in Single-Crystal Structure Analysis. In *Crystallographic Computing*, Ahmed, F. R., Ed. 1970; pp 255-270.
34. *CrystalClear*, Rigaku Corporation: Tokyo, Japan, 2015.
35. *CrystalStructure*, 4.2.4; Rigaku Corporation: Tokyo, Japan, 2016.
36. Spek, A. L. *J. Appl. Cryst.* **2003**, *36*, 7-13.
37. Brandenburg, K. *DIAMOND*, Crystal Impact GbR: Bonn, Germany, 2017.

Appendix

D.4 Table of Compounds

1a	[^{C2}(OSSO)^{tBu,tBu}MoCl$_2$]
1b	[^{Cy}(OSSO)^{CPhMe2,CPhMe2}MoCl$_2$]
2a	[^{C2}(OSSO)^{tBu,tBu}MoOCl]
2b	[^{C2}(OSSO)^{tBu,pMe}MoOCl]
2c	[^{Cy}(OSSO)^{tBu,tBu}MoOCl]
3	[^{C2}(OSSO)^{tBu,tBu}MoO$_2$]
4	[^{Cy}(OSSO)^{tBu,tBu}WOCl]
[5$_{Ca}$][(OAc)$_2$]	[^{C3}LZn$_3$Ca(OAc)$_2$]
[5$_{La}$][(OAc)$_3$]	[^{C3}LZn$_3$La(OAc)$_3$]
[6$_{Ca}$][(OTf)$_2$]	[H$_6$^{Cy}LCa(OTf)$_2$]
[7$_{Ca}$][(OTf)$_2$]	(R,R)-[H$_6$^{Cy}LCa(OTf)$_2$]
[7$_{Sr}$][(OTf)$_2$]	(R,R)-[H$_6$^{Cy}LSr(OTf)$_2$]
[7$_{Ba}$][(OTf)$_2$]	(R,R)-[H$_6$^{Cy}LBa(OTf)$_2$]
[7$_{La}$][(OTf)$_3$]	(R,R)-[H$_6$^{Cy}LLa(OTf)$_3$]
[7$_{Ce}$][Cl$_3$]	(R,R)-[H$_6$^{Cy}LCeCl$_3$]
[8$_{Ca}$][(OTf)$_2$]	(S,S)-[H$_6$^{Cy}LCa(OTf)$_2$]
[9$_{Ca}$][(OTf)$_2$]	(R,R)-[H$_6$^{C2Ph2}LCa(OTf)$_2$]
[9$_{Ca}$][Cl$_2$]	(R,R)-[H$_6$^{C2Ph2}LCaCl$_2$]
[9$_{Ca}$][Br$_2$]	(R,R)-[H$_6$^{C2Ph2}LCaBr$_2$]
[9$_{Ca}$][I$_2$]	(R,R)-[H$_6$^{C2Ph2}LCaI$_2$]
[9$_{Sr}$][(OTf)$_2$]	(R,R)-[H$_6$^{C2Ph2}LSr(OTf)$_2$]
[9$_{Ba}$][(OTf)$_2$]	(R,R)-[H$_6$^{C2Ph2}LBa(OTf)$_2$]
[9$_{Ba}$][(OTs)$_2$]	(R,R)-[H$_6$^{C2Ph2}LBa(OTs)$_2$]
[9$_{La}$][(OTf)$_3$]	(R,R)-[H$_6$^{C2Ph2}LLa(OTf)$_3$]
[9$_{Ce}$][Cl$_3$]	(R,R)-[H$_6$^{C2Ph2}LCeCl$_3$]
10	[H$_6$^{Cy}L]
[11$_{Ca}$][I$_2$]	[((R,R)-^{C2Ph2}L)Zn$_3$CaI$_2$]
[11$_{Ca}$][(OAc)$_2$]	[((R,R)-^{C2Ph2}L)Zn$_3$Ca(OAc)$_2$]
[11$_{Ce}$][Cl$_3$]	[((R,R)-^{C2Ph2}L)Zn$_3$CeCl$_3$]
[12$_{Ca}$][(OTf)$_2$]	[((R,R)-^{Cy}L)(VO)$_3$Ca(OTf)$_2$]
[13$_{Ca}$][(OTf)$_2$]	[((R,R)-^{C2Ph2}L)(VO)$_3$Ca(OTf)$_2$]
[13$_{Ca}$][Cl$_2$]	[((R,R)-^{C2Ph2}L)(VO)$_3$CaCl$_2$]
[13$_{Ca}$][Br$_2$]	[((R,R)-^{C2Ph2}L)(VO)$_3$CaBr$_2$]
[13$_{Ca}$][I$_2$]	[((R,R)-^{C2Ph2}L)(VO)$_3$CaI$_2$]
[13$_{Sr}$][(OTf)$_2$]	[((R,R)-^{C2Ph2}L)(VO)$_3$Sr(OTf)$_2$]
[13$_{Sr}$][I$_2$]	[((R,R)-^{C2Ph2}L)(VO)$_3$SrI$_2$]

Appendix

[13$_{Ba}$][(OTs)$_2$]   [((R,R)-^{C2Ph2}L)(VO)$_3$Ba(OTs)$_2$]
[13$_{Ba}$][I$_2$]   [((R,R)-^{C2Ph2}L)(VO)$_3$BaI$_2$]
[13$_{Ce}$][Cl$_3$]   [((R,R)-^{C2Ph2}L)(VO)$_3$CeCl$_3$]
[13$_{La}$][(OTf)$_3$]   [((R,R)-^{C2Ph2}L)(VO)$_3$La(OTf)$_3$]
[14$_{Ca}$][(OTf)$_2$]   [((R,R)-^{C2Ph2}L)(TiO)$_3$Ca(OTf)$_2$]
[15$_{Ca}$][(OTf)$_2$]   [((R,R)-^{C2Ph2}L)Li$_6$Ca(OTf)$_2$]
[16$_{Ca}$][Cl$_2$]   [((R,R)-^{C2Ph2}L)Mn$_3$CaCl$_2$]

Appendix

D.1 Eidesstattliche Erklärung

Ich, Tobias Schindler,

erkläre hiermit, dass diese Dissertation und die darin dargelegten Inhalte die eigenen sind und selbstständig, als Ergebnis der eigenen originären Forschung, generiert wurden.

Hiermit erkläre ich an Eides statt:

1. Diese Arbeit wurde vollständig oder größtenteils in der Phase als Doktorand dieser Fakultät und Universität angefertigt;

2. Sofern irgendein Bestandteil dieser Dissertation zuvor für einen akademischen Abschluss oder eine andere Qualifikation an dieser oder einer anderen Institution verwendet wurde, wurde dies klar angezeigt;

3. Wenn immer andere eigene oder Veröffentlichungen Dritter herangezogen wurden, wurden diese klar benannt;

4. Wenn aus anderen eigenen oder Veröffentlichungen Dritter zitiert wurde, wurde stets die Quelle hierfür angegeben. Diese Dissertation ist vollständig meine eigene Arbeit, mit der Ausnahme solcher Zitate;

5. Alle wesentlichen Quellen von Unterstützung wurden benannt;

6. Wenn immer ein Teil dieser Dissertation auf der Zusammenarbeit mit anderen basiert, wurde von mir klar gekennzeichnet, was von anderen und was von mir selbst erarbeitet wurde;

7. Ein Teil oder Teile dieser Arbeit wurden zuvor veröffentlicht und zwar in:
   (1) Schindler, T.; Sauer, A.; Spaniol, T. P.; Okuda, J. Oxygen Atom Transfer Reactions with Molybdenum Cofactor Model Complexes That Contain a Tetradentate OSSO-Type Bis(phenolato) Ligand. *Organometallics*, **2018**, *37*, 4336–4340.

Datum:

Unterschrift:

D.2 Curriculum Vitae

Personal Information

Name	Tobias Schindler
Address	Rudolfstr. 40
	52070 Aachen
	Germany
Phone	+49-157-71 90 03 23
E-mail	tobias.schindler@rwth-aachen.de
Date of Birth	February 4, 1991
Place of Birth	Goslar, Germany

Internships and Working Experience

Since April 2019	**Scientist Catalysis Research**
	ARLANXEO Netherlands B.V., Geleen, Netherlands
05/2016 – 03/2019	**Scientific Assistant**
	Institute of Inorganic Chemistry, RWTH Aachen University
09/2017 – 12/2017	**Short Term Visiting Research Scientist, Japan**
	Graduate School of Engineering Science, Osaka University, Japan
10/2015 – 03/2016	**Student Science Assistant**
	Institute of Inorganic Chemistry, RWTH Aachen University
06/2015 – 07/2015	**Research Intern**
	Max-Planck-Institute of Coal Research
10/2014 – 03/2015	**Student Science Assistant**
	CAT Catalytic Center at RWTH Aachen University
10/2013 – 02/2014	**Intern at the BASF R&D Department, Shanghai**
	BASF Auxiliary Chemicals Co. Ltd., Shanghai, P. R. China
10/2008 – 10/2008	**Internship as Lab Assistant**
	Institute of Organic Chemistry, TU Clausthal

Education

Since May 2016	**Doctoral Student**
	Research group of Prof. Dr. Jun Okuda, RWTH Aachen University
10/2010 – 03/2016	**Graduate and Undergraduate Studies in Chemistry**
	RWTH Aachen University, Germany
Specialization	➢ Specialization in catalysis and mesoscopic systems
Degrees	➢ Bachelor and Master studies completed within standard period of studies: M. Sc., grade: 1.4 (very good); B. Sc., grade: 1.9 (good)
2010	**Abitur**
	Robert-Koch-Schule, Clausthal-Zellerfeld, Germany
08/2007 – 06/2008	**High School**
	Barnum Public High School, Minnesota, USA
	➢ Foreign exchange student with international Experience e.V.

Appendix

D.3 Table of Publications

D.3.1 Peer-Reviewed Publications

(1) Schindler, T.; Sauer, A.; Spaniol, T. P.; Okuda, J. Oxygen Atom Transfer Reactions with Molybdenum Cofactor Model Complexes That Contain a Tetradentate OSSO-Type Bis(phenolato) Ligand. *Organometallics* **2018**, *37*, 4336–4340.

(2) Riensch, N. A.; Fritze, L.; Schindler, T.; Kremer, M.; Helten, H. Difuryl(supermesityl)borane: a versatile building block for extended π-conjugated materials. *Dalton Trans.* **2018**, *47*, 10399-10403.

(3) Nagae, H.; Aoki, R.; Akutagawa, S.; Kleemann, J.; Tagawa, R.; Schindler, T.; Choi, G.; Spaniol, T. P.; Tsurugi, H.; Okuda, J.; Mashima, K. Lanthanide Complexes Supported by a Trizinc Crown Ether as Catalysts for Alternating Copolymerization of Epoxide and CO_2: Telomerization Controlled by Carboxylate Anions. *Angew. Chem. Int. Ed.* **2018**, *57*, 2492-2496.

(4) Schindler, T.; Paparo, A.; Nishiyama, H.; Spaniol, T. P.; Tsurugi, H.; Mashima, K.; Okuda, J. Deprotonation of a formato ligand by a *cis*-coordinated carbyne ligand within a bis(phenolate) tungsten complex. *Dalton Trans.* **2018**, *47*, 13328-13331.

(5) Schindler, T.; Lux, M.; Peters, M.; Scharf, L.; Osseili, H.; Maron, L.; Tauchert, M. E. Synthesis and Reactivity of Palladium Complexes Featuring a Diphosphinoborane Ligand. *Organometallics* **2015**, *34*, 1978-1984.

D.3.2 Conference Contributions

(1) Oral Presentation: Mono- and Mixed Mutlimetallic Complexes Featuring [3+3]-Macrocyclic Ligands and Their Application in Catalysis. At *Green Growth*, Osaka, Japan, 2018.

(2) Oral Presentation: Group 6 Metal Complexes Featuring Tetradentate Bis(phenolate) Ligands. At *Green Growth*, Osaka, Japan, 2017.

(3) Oral Presentation: Group 6 Metal Complexes Featuring Tetradentate Bis(phenolate) Ligands. At *Workshop on Artificial Metalloenzymes and Joint SeleCa Symposium*, Aachen, Germany, 2016.

(4) Poster Presentation: Mono- and Mixed Mutlimetallic Complexes Featuring [3+3]-Macrocyclic Ligands and Their Application in Catalysis. At *SeleCa Symposium*, Aachen, Germany, 2018.

(5) Poster Presentation: Decomposition of a Tungsten Propylidyne Complex Featuring a *cis*-coordinated Formato Ligand. At *SeleCa Symposium*, Aachen, Germany, 2017.

(6) Poster Presentation: Decomposition of a Tungsten Propylidyne Complex Featuring a *cis*-coordinated Formato Ligand. At *EuCOMC*, Amsterdam, Netherlands, 2017.

Appendix

D.3.3 Other Publications

(1) Schindler, T.; Spaniol, T. P; Tauchert, M. E. Abschlusssymposium des Graduiertenkollegs SeleCa. *Nachr. Chem.* **2019**, *67*, 79.

(2) Sauer, D. F.; Schindler, T. Deutsch-Japanischer Katalyse-Workshop. *Nachr. Chem.* **2017**, *65*, 73.

Appendix

D.4 Acknowledgements – Danksagung

An erster Stelle möchte ich mich bei meinem Doktorvater Univ.-Prof. Dr. Jun Okuda für die Aufnahme in seinen Arbeitskreis, die hervorragenden Arbeitsbedingungen, sowie für die wissenschaftliche Betreuung bedanken. Insbesondere danke ich ihm dafür, dass er mich dazu ermutigt hat, mich für das internationale Graduiertenkolleg SeleCa zu bewerben. Hierdurch hatte ich die Möglichkeit, einen Teil meiner Doktorarbeit in Japan anzufertigen.

I would like to thank our collaborators Prof. Kazushi Mashima, Assoc. Prof. Hayato Tsurugi, Ass. Prof. Haruki Nagae and the other members of the Mashima Lab for their great support during my stay in Japan. I want to thank Prof. Kazushi Mashima for being my second examiner and for giving me the opportunity to work on the very exciting and challenging macrocycle topic. Especially, I would like to express my deepest gratitude to Ass. Prof. Haruki Nagae for many fruitful discussions, for a good collaboration and for his great support during my stay in Japan.

Ich möchte mich herzlichst bei Univ.-Prof. Dr. Sonja Herres-Pawlis für die Übernahme der Funktion als Drittprüferin bedanken. Darüber hinaus bedanke ich mich für die Möglichkeit, EPR Spektren in ihrem Arbeitskreis messen lassen zu dürfen. Ich bedanke mich bei Univ.-Prof. Dr. Markus Albrecht für die Übernahme der Funktion des Prüfungsvorsitzenden.

Ich bedanke mich bei allen aktuellen und ehemaligen Mitgliedern des Arbeitskreises Okuda. Zudem möchte ich mich bei Dr. Daniel F. Sauer und bei Dr. Klaus Beckerle für die vielen anregenden Diskussionen und Hilfestellungen während meiner Promotionszeit bedanken. Außerdem möchte ich mich sehr bei Dr. Michael Tauchert für seine Ratschläge bedanken – ohne ihn wäre meine Doktorarbeit anders verlaufen. Mein besonderer Dank geht auch an Dr. Thomas Spaniol, Dr. Daniel F. Sauer, Andreas Thiel und Dr. Priyabrata Ghana für die Anmerkungen und Korrekturvorschläge des Manuskriptes. Ein großer Dank geht auch an Simone Becher für die Unterstützung bei allen organisatorischen Fragen und für die gute Zusammenarbeit im SeleCa Graduiertenkolleg.

Kristina Keisers danke ich für die vielen EPR Messungen. Dem Arbeitskreis von Univ.-Prof. Dr. Paul Kögerler, Dr. Dr. Jan van Leusen und Christina Houben danke ich für die SQUID Messungen und Simulationen. Dr. Gerhard Fink, Rachida Bohmarat und Toni Gossen danke ich für die Unterstützung bei den NMR Messungen. Florian Ritter, Ass.-Prof. Haruki Nagae, dem AK Englert und Dr. Thomas Spaniol danke ich für die Röntgenstrukturanalyse. Brigitte Pütz danke dich für die zahlreichen massenspektrometrischen Messungen. Frau Schleep danke ich für die Elementaranalysen. Monika Paul danke ich für die GPC Messungen. Ich danke allen Mitarbeiter des Instituts für Anorganische Chemie für die Unterstützung und gute Zusammenarbeit über die Jahre.

Appendix

Ein großer Dank geht auch an die SeleCats für eine wirklich großartige Zeit im Graduiertenkolleg und während unserer Aufenthalte in Japan.

Ein großer Dank gilt meiner Familie, meinen Freunden und insbesondere meinen Eltern, die mich während meines Studiums und meiner Doktorarbeit immer unterstützt und an mich geglaubt haben.

Mein größter Dank geht an meinen Freund Tobias Schreckenbach. Du bist immer für mich da und hast mich während meiner Promotionszeit immer wieder aufgebaut. Zusammen haben wir diese schwierige Zeit geschafft und dafür bin ich dir zutiefst dankbar.